FREE BOOKS

www.forgottenbooks.org

You can read literally <u>thousands</u> of books
for free at www.forgottenbooks.org

(please support us by visiting our web site)

Truth may seem, but cannot be:
Beauty brag, but 'tis not she;
Truth and beauty buried be.

To this urn let those repair
That are either true or fair;
For these dead birds sigh a prayer.

Bacon

A Treatise on
The Astrolabe

addressed to his son Lowys

by

Geoffrey Chaucer

A.D. 1391

EDITED FROM THE EARLIEST MSS.

BY

WALTER W. SKEAT

'His Astrelabie, longynge for his art'

Canterbury Tales, A 3209

Published for

THE EARLY ENGLISH TEXT SOCIETY

by the

OXFORD UNIVERSITY PRESS

LONDON   NEW YORK   TORONTO

FIRST PUBLISHED 1872
REPRINTED 1880, 1928, 1968

Extra Series, No. 16

ORIGINALLY PRINTED BY
RICHARD CLAY & SONS LTD., BUNGAY, SUFFOLK
AND NOW REPRINTED LITHOGRAPHICALLY IN GREAT BRITAIN
AT THE UNIVERSITY PRESS, OXFORD
BY VIVIAN RIDLER, PRINTER TO THE UNIVERSITY

CONTENTS.

PREFACE.

§ 1. The existing MSS. of the "Astrolabe" are still numerous. I have been successful in finding no less than *eighteen*, sixteen of which I here describe.[1] It is remarkable that, although many printed editions of the treatise have appeared, no first-class MS. has ever hitherto come under the notice of any one of the various editors. This point will appear more clearly hereafter.

§ 2. A.—MS. Dd. 3. 53 (part 2) in the Cambridge University Library. The "Treatise on the Astrolabie" begins at fol. 212 of the MS. considered as a whole, but the folios are now properly renumbered throughout the treatise, as in the present volume. The MS. is of vellum, and the writing clear and good, with a great number of neatly drawn diagrams, which appear wherever the words "lo here thi figure" occur in the text. This MS. I have made the basis of the text, and it is followed with minute exactness except when notice to the contrary is given in the Critical Notes. Wherever any change of even slight importance is made, notice is drawn to the alteration by the use of square brackets.

This MS. is of considerable importance. The hand-writing

---

[1] Two were kindly pointed out to me by Mr Bradshaw after this Preface was in type. Both are imperfect. They are (1) MS. Bodley 68, ending with Part ii. sect. 36, chiefly remarkable for containing the title "Bred and Mylk for children"; and (2) MS. E Museo 116, in the Bodleian Library, which contains a fragment of the latter part of the treatise on vellum, in the hand-writing of the scribe of MS. Camb. Gg. 4. 27.

exactly resembles that in MS. B., and a comparison of the MSS. leads to the following results. It appears that MSS. A. and B. were written out by the same scribe, nearly at the same time. The peculiarities of spelling, particularly those which are faulty, are the same in both in a great many instances. It is also clear that the said scribe had but a very dim notion of what he was writing, and committed just such blunders as are described in Chaucer's Lines to Adam Scrivener, and are there attributed to "negligence and rape."[1] It is still more interesting to observe that Chaucer tells us that he had to amend his MSS. by "rubbing and scraping" with his own hand; for MS. A. and B. differ precisely in this point, viz. that while the latter is left uncorrected, the former has been diligently "rubbed and scraped" by the hand of a corrector who well knew what he was doing, and the right letters have been inserted in the right places over the erasures. These inserted letters are in the hand of a second scribe who was a better writer than the first, and who was entrusted with the task of drawing the diagrams. The two hands are contemporaneous, as appears from the additions to the diagrams made by the writer of the text. Unfortunately, there are still a good many errors left. This is because the blunders were so numerous as to beguile the corrector into passing over some of them. When, for example, the scribe, having to write "lo here thi figure" at the end of nearly every section, took the trouble to write the last word "vigure" or "vigour" in nearly every instance, we are not surprised to find that, in a few places, the word has escaped correction. It further appears that some of the later sections, particularly sections 39 and 40, have not been properly revised; the corrector may very well have become a little tired of his task by the time he arrived at them. It must also be remembered, that such blunders as are made by a scribe who is not clear as to the meaning of his subject-matter are by no means the blunders which are most puzzling or most misleading; they are obvious at once as evident blotches, and the general impression left upon the mind by the perusal of this MS. is—that a careless scribe copied it from some almost perfect original, and that his errors were

---

[1] I. e. haste, rapidity. Cf. "Rydynge ful *rapely;*" Piers the Plowman, B. xvii. 49.

partially corrected by an intelligent corrector, who grew tired of his task just towards the end.

The order of the conclusions in Part ii. differs from that in all the editions hitherto printed, and the MS. terminates abruptly in the middle of a sentence, at the words "howre after howre" in Conclusion 40. A portion of the page of the MS. below these words is left blank, though the colophon "Explicit tractatus," &c., was added at the bottom of the page at a later period.

Certain allusions in the former part of the MS. render it probable that it was written in London, about the year 1400.

§ 3. B.—MS. E Museo 54, in the Bodleian Library, Oxford. This is an uncorrected duplicate of the preceding, as has been explained, and ends in the same way, at the words "howre after howre," followed by a blank space. The chief addition is the rubricated title—"Bred and mylk For childeren," boldly written at the beginning; in the margin are the following notes in a late hand —"Sir Jiffray Chaucer"—"Dominus Gaufredus Chaucerus"— "Galfredi Chauceri Tractatus de Ratione et vsu Astrolabij ad Ludouicum filium." At the end is the note—"Liber Francisci Beyley, 1637. Franc. Bayley, Noui Collegij Socius, Anno Dom., 1637. Ned. Tourner."

Before I undertook the present edition, a transcript of part of this MS. had been made for the Early English Text Society, which afterwards came into my hands. A portion of the text was "set up" from it, but the proof sheets were corrected by MS. A. I mention this to show how *closely* the two MSS. resemble each other in spelling. It is very seldom that such a course is practicable; but in this instance it occasioned no difficulty.

§ 4. C.—MS. Rawlinson, Misc. 1370 (leaves 22—42), in the Bodleian Library, Oxford. This is a beautifully written MS., on vellum, with 38 pages of text, and 4 blank pages. It has the conclusions in the same order as the preceding, six well-executed diagrams, and corrections on nearly every page. It is of early date, perhaps about A.D. 1420, and of considerable importance. It agrees closely with the text, and, like it, ends with "howre after howre." Some variations of spelling are to be found in the Critical Notes. In this MS. the "Conclusions" are numbered in the margin, and the numbers agree with those adopted in this edition.

§ 5. D.—MS. Ashmole 391, in the Bodleian Library. This contains several tracts of very different dates—including tracts on astrology, calendars, tables, a *printed* tract, a tract on houses and horoscopes, a Latin tract with a very carefully painted picture resembling that given as fig. 19 in this volume, and finally, Chaucer's "Astrelabie." This is an old and well-written copy on vellum, with illuminated border on the first page, fair diagrams, blue and flourished capital letters, &c., and is much faded. It begins—"Lite lowys my sone, I aparceyue wel by certeyn euydences"—and contains the following, viz. all of Part i; Part ii, sections 1, 2, and part of 3, down to "18 degrees of heighte taken bi myn" in l. 30, after which several leaves are lost; then comes sect. 25, beginning at l. 17—"but for ensaumple; For wel I woot þe latitude of Oxenford," &c., followed by sections 26, 27, 28, 29, 30, and part of 31, down to l. 9 —"The maner of diuysion of þe." The rest is torn away. I have made but little use of this MS., on account of its being so imperfect.

§ 6. E.—MS. Bodley 619. This MS., like B., has the title— "Brede and Milke for children." Like other good MSS., it ends sect. 40 with "houre after houre." But after this, there occurs an additional section, which is probably not genuine, but which I have printed here (for the sake of completeness) as section 46; which see.

There are some Latin notes in this MS. which are worth notice. The first is a note on Chaucer's words in Part i, sect. 10, l. 14, that "the sonne dwelleth ther-for neuere the more ne lesse in on signe than in another," which declares this to be a mistake, for the sun dwells longer in Cancer than in Capricorn; an observation which is perfectly correct.

Again, at the end of sect. 3 in Part ii, we have a Latin paragraph, beginning—"Nota, quod si quot miliaria sunt inter duas regiones"—and ending—"dando 100. miliaria. Idem facies de longitudinibus, si fuerint diuerse, & latitudines eodem." This is a quotation from Messahala (see p. 97), and is very interesting, because it directly connects Chaucer's translation with the Latin text of Messahala.

At fol. 53, back, we find another Latin note, having reference to Part ii, sect. 39, as follows :—

" Nota ; si vis scire per quot gradus currit Almicantatium, computa almicantarath, incipiendo ab orisonte vsque ad Cenith, et per numerum illorum diuide 90, et numerus quociens ostendet tibi per quot currat.

" Longitudines autem quarundam regionum, idem elongaciones circulorum earum meridianorum a meridiano vltime regionis habitabilis in occidente. Et earum latitudines, idem distancias ab equinoxiali circulo, notabimus in quadam tabula."

This is of some interest, as shewing that the ancients took for their first meridian of longitude the meridian of the last habitable spot which could be reached in proceeding westward. The principle is clear, but the locality vague. Observe that the latter part of this note is also from Messahala; see p. 97.

At fol. 15, there is a note on Part i, sect. 21, l. 12, where Chaucer instances the stars Aldebaran and Algomeysa. To these are here added the stars " Menkar," " Algevze," and " cor leonis," that is to say, α Ceti, α Orionis, and α Leonis ; with the remark— " nota : þat þese 5 sterres ben meridional fro þe ecliptic, and septentrional fro þe equinoctial, secundum astrolabium colleg. de Merton." Merton College, it thus appears, possessed an Astrolabe on which the five above-named stars were represented.

At fol. 21 is an additional section, not found elsewhere, which is printed in the Additional Notes ; see p. 81. This conclusion has some claims to our notice, because, whether genuine or not, it is translated from Messahala.

§ 7. F.—MS. 424, in the library of Corpus Christi College, Cambridge. Very imperfect, especially at the beginning, where a large portion has been lost. Written in a close hand, late in the fifteenth century, though the thorn-letter (þ) appears in it. Begins— " vnderstond well þat þe zodiake is departyd in 2 halfe cercles as fro þe hede of capricorne "—which is sect. 16 of Part ii without the rubric. Then follow, with rubrics, the entire sections 17—36, the

last of which ends thus, with an additional remark—" & the begynnyng of þe 12 howse is nadyr to þe 6. ¶ To fynde þe howse by þe
astrolaby þat is wretin[1] suffyse. Explicit *tractatus* astrolabii *secundum* chausers, *factus* filio suo lodowyco."

Although the MS. is thus imperfect, we see that the conclusions
follow the right order, as in the best MSS.

§ 8. G.—MS. R. 15. 18, in the library of Trinity College, Cambridge. This is a curious and interesting volume, as it contains
several tracts in English on astrology and astronomy, with tables of
stars, &c. It also contains the picture which I have but imperfectly
represented in Fig. 19.[2]

The copy of the "Astrolabe" in this MS. is not a very good one.
It is not divided into paragraphs or sections, and occasionally portions of sentences are omitted. It ends with the words—" as well
as by the fyxe sterre" in Part ii, sect. 34, 1. 14. The conclusions
are in the right order, and there are a few diagrams.

§ 9. H.—MS. Sloane 314, British Museum. A late MS. on
paper, absurdly said in a note to be in Chaucer's handwriting,
whereas it is clearly to be referred to the end of the fifteenth century.
Size of page, about 8 inches by 5½. The treatise begins on fol. 65,
back, and ends on fol. 106, in the middle of a page, at the end of
conclusion 36, like MS. F. It is written in a clear hand, but with
pale ink. It has rubrics in red, and some not very well-drawn
diagrams. The conclusions are (unless I have misread my notes) in
the wrong order, i. e. in the order adopted in the old printed editions.

§ 10. I.—MS. Sloane 261. This is an "edited" MS., having
been apparently prepared with a view to publication. Mr Brae
has made considerable use of it, and gives, in his preface, a careful
and interesting account of it. He concludes that this MS. was
written by Walter Stevins in 1555, and dedicated by him to Edward
Earl of Devonshire ; and that MS. H. was one of those which Stevins
especially consulted, because it contains marginal notes in Stevins'

---

[1] Very indistinct. MS. Addit. 23002 has "ywrytten" here.

[2] I regret to say that my hasty copy of this picture gives merely the general idea of it. The truth is, I was not aware of the marvellous accuracy with
which such a wood-engraver as Mr Rimbault can reproduce what is given him,
or I would have taken care to copy it more exactly.

handwriting. The date 1555 was assigned to it by Mr Brae after most careful investigation; in any case, it is the latest MS. which I know of. A memorandum shews that this MS. was in Urry's hands in 1712; a fact which is (as Mr Brae points out) not much to Urry's credit, seeing that some of the glaring errors in Urry's edition might have been corrected by consulting Stevins. The contents of this MS. can be so well ascertained from Mr Brae's edition that it is unnecessary to say more about it here. The Conclusions are arranged in the same order as in other MSS. not of the first class. This will be further discussed presently.

§ 11. K.—MS. Rawlinson Misc. 3, in the Bodleian Library, Oxford. On vellum, 49 folios, with rich gold capitals, beautifully ornamented; in a large clear handwriting, with red rubrics. Title—"Astralabium." Begins—"Lityl lowys my sone," &c.—and ends—"For þe mone meuyth the contrarie from other planetys. as yn here epicircle. but in none other maner;" see end of Part ii, sect. 35. Order of Conclusions in Part ii as follows; 1—12, 19—21, 13—18, 22—35; as in other late MSS. There are no diagrams, and the MS., though well written, may perhaps be referred to the latter half of the fifteenth century.

§ 12. L.—MS. Additional 23002, British Museum. A fair MS., on vellum, without diagrams; size of page, about 7½ by 5 inches. Begins on fol. 3; ends on fol. 28b. Contents as follows:—Part i, wanting sections 15—23 inclusive; Part ii, sections 1—12, 19—21, 13—18, 22—35, as in K.; together with additional sections, viz. 41—43; also 44, 45; also 41a—42b; then 36 and 37, concluding with the words "of 3 howses that folowyn." The second part is thus seen to be nearly complete, although sections 38—40 are missing. See also the Additional Note on Part ii, sect. 3.

§ 13. M.—MS. E. 2 in the library of St John's College, Cambridge. Small MS. on vellum, without diagrams. Size of page, 6 in. by 4 in. Former owner, Wilielmus Graye. Contents: (a) Fol. 1. De septem climatibus expositio (Short treatise in Latin); (b) Fol. 2. De astrolabio. The leaves have been misplaced, and bound up in a wrong order, but nothing is lost. If they were properly rearranged, the order of contents of Part ii would be seen to be as follows, viz.

sections 1—12, 19—21, 13—18, 22—35, as in the last MS.; with
the additional sections 41—43; also 44, *but not* 45; also 41*a*—42*b*;
after which come sections 36—38, the last ending with the words
"styke an euen pyn or a were vpriȝt, þe smallere þe bettre.   sette þy
pyn be plum-rewle euen"; see l. 6.   I have printed from this MS.
the last five words of sect. 40; also 41—43, and 41*a*—42*b*; besides
collating it for the improvement of the text in sect. 44.   I have also
been indebted to it for the *Latin* rubrics to the conclusions, which I
have not found elsewhere.   Several various readings from this MS.
appear in the Critical Notes.

§ 14. N.—MS. Digby 72, in the Bodleian Library.   This is a
collection of various tracts, including tables of latitudes of planets, and
for finding the moon's place; table of roots of " mene motes " for the
" anni collecti," &c. (see Part ii, sect. 44); tables of the motions of
the sun and moon; astrological tables; description of planets; on
horoscopes; on aspects; after which, on fol. 78, a curious table of 15
fixed stars, in which each star is denoted by some odd-shaped strag-
gling character, and is connected with certain gems and herbs.   On fol.
79 comes the " Astrolabye," beginning—" lytull lewis my zone, I
perseyve well," &c.   The conclusions in Part ii are : 1—12, 19—21,
13—18, 22—35; 41—43; 44, 45; 41*a*—42*b*; 36 and 37, ending
with the words—" 3 howsis that folowen;" cf. MS. L.   From this
MS. I have printed the text of sections 44 and 45, but have made little
further use of it.   The writing is not very good, and the ink pale.

§ 15. O.—MS. Ashmole 360, in the Bodleian Library.   Late MS.,
on paper; former owner's name, Johan Pekeryng; without diagrams.
There are evidently some omissions in it.   But it includes sections 44
and 45, and I have given various readings from it in those sections.   It
ends at the end of sect. 43*a*, with the words—"*one* to twelfe. *& sic finis.*"

§ 16. P.—MS. Dd. 12. 51 in the Cambridge University Library.
Small MS. on vellum, size of pages scarcely 6 inches by 4; containing
86 leaves, and written in the fifteenth century.   The text is by no
means a bad one, though the spelling is somewhat peculiar.   Unfortunately, some of the pages are very much rubbed and defaced; otherwise I should have made more use of it.   As it is, I have taken from
it some various readings, recorded in the Critical Notes.   The scribe

seems generally to have understood what he was writing, which is not often the case in MSS. of the "Astrolabe;" so that this MS. is useful in passages where other texts have absurd readings.

One point deserves particular attention. It not only contains the conclusions of Part ii *in the right order*, but continues it *without a break* to the end of conclusion 43; at the end of which is the colophon—Explicit tractatus astrolabii.[1]

§ 17. Q.—MS. Ashmole 393, in the Bodleian Library; on paper. This is of little importance. The piece entitled "Chauucers: The Tretyse off the Astrolabye" merely fills one closely-written leaf, and contains a sort of epitome of Part i, with the beginning of Part ii.

§ 18. Of the above MSS., Mr Brae describes H., I., and L. only, and does not seem to have made use of any others. Mr Todd, in his Animadversions on Gower and Chaucer, p. 125, enumerates only four MSS., which are plainly A., P., F., and G. The rest seem to have escaped attention.

In addition to the MS. authorities, we have one more source of text, viz. the Editio Princeps, which may be thus described.

R.—The edition of Chaucer's Works by Wm. Thynne, printed at London by Thomas Godfray in 1532. This is the first edition in which the Treatise on the Astrolabe appeared; it begins at fol. ccxcviii., back. The Conclusions in Part ii are in the order following, viz. 1—12, 19—21, 13—18, 22—40; after which come 41—43, and 41*a*—42*b*. This order does not agree precisely with that in any MS. now extant, with the exception of I., which imitates it. It is further remarkable for certain additions and errors, which are discussed in § 26 below. All later editions, down to Urry's in 1721, contribute no new information. The few slight alterations which appear in them are such as could have been made without reference to MSS. at all.

REMARKS ON THE CLASSES OF THE MSS.

§ 19. On comparing the MSS., it at once appears that they do not agree as to the order of the Conclusions in Part ii. The MSS. A., B.,

---

[1] This MS. is, in fact, of the first class, and should have been mentioned much earlier; but the mistake was overlooked till it was too late to correct it.

C. (which are unquestionably the oldest) as well as E., F., G., and P.,
adopt the order which appears in this edition, but which has never
appeared in any previous edition.   In all other editions we find the
three sections 19—21 made to precede sections 13—18.   Now we might
here appeal to authority only, and say that the order in the *oldest*
MSS. ought to be preferred.   But it so happens that we can appeal
to internal evidence as well, and there are at least three considerations
which shew that the oldest MSS. are certainly correct.   These are as
follows.   In the *first* place, sect. 18 amounts to finding the degree of
the zodiac which *souths* with any star, and begins with the words
" Set the centre of the sterre vpon the lyne Meridional " ; whilst sect.
19 amounts to finding the degree of the zodiac that *rises* with any
star, and begins with the words " Set the sentre of the sterre vpon the
est orisonte."   Clearly, these " conclusions" are closely linked together, and one ought to follow the other.   But, in all the editions,
this continuity is broken.   In the *second* place, the rubric of sect. 21
is—" To knowe for what latitude in any regioun," &c. ; whilst that
of sect. 22 is—" To knowe in special the latitude of owre countray,"
&c.   Clearly, these conclusions are closely linked, and in their right
order.   But, in all the editions, this continuity is again broken ; and
we have this absurd result, viz. that a proposition headed—" To
knowe the degrees of the longitudes of fixe sterres " is followed by one
headed—" To knowe *in special* the latitude of owre countray."
What in the world can the latitude of a place have to do with
the longitude of a star?   And how is it possible to assign, in this arrangement, the faintest idea of sense to the words "*in special* " ?   This
argument is alone convincing.   But *thirdly*, we may note the heading
of sect. 16—" This chapitre is a Maner declaracioun to conclusiouns
þat folwen."   By the right arrangement, this section comes earlier
than it does otherwise, and precedes sections 19, 20, and 21, which is
a more natural arrangement than that in former editions.   This is a
minor point, and I lay no stress on it.   But the two former reasons
are cogent, and we see that common sense confirms that arrangement
of sections which the authority of the oldest MSS. prescribes.   The
two things together are sufficient, and we can now trust to the oldest
MSS. with the greater confidence.   Hence we are enabled to draw a

line, and to divide the MSS. into two classes; those in which the order of sections is correct, and those in which it has suffered misplacement, the number in each class being much the same. This gives us the following result.

First Class. A. B. C. (probably D.) E. F. G. P.

Second Class. H. I. K. L. M. N. O.; to which add R.

But this division immediately leads to another very curious result, and that is, a certain lack of authority for sections after the *fortieth.*

A. ends with an incomplete sentence, in sect. 40, with the words howre after howre."

B. C. end exactly at the same place.

E. ends sect. 40 with the same words; and, after this, has only one additional section (46), which is, in my opinion, spurious; especially as it does not appear in Messahala, of which more anon.

D. fails earlier, viz. in sect. 31, which is incomplete.

F. has all down to the end of sect. 36, and then—" explicit."

G. breaks off in sect. 34, which is incomplete.

In none of the first-class MSS. (excepting P., which terminates with section 43) is there a word about *umbra recta* or *umbra versa.*

Even in the second class of MSS., we find H. breaking off at sect. 36, and K. at sect. 35; so that the sections on the *umbrœ* rest only on MSS. I. (obviously an edition, not a transcript), L., M., N., O., and P. Putting aside the first of these, as being "edited," we have but five left; and in the first four of these we find that the additional Conclusions appear in a certain order, viz. they insert 44 and 45 (on the "mene mote") between three sections 41—43 on the "umbræ" and five other sections 41*a*—42*b* on the same.

§ 20. This at once suggests two results. The *first* is, that, as this gives two sets of sections on the "umbræ," we can hardly expect both to be genuine; and accordingly, we at once find that the *last five* of these are mere clumsy repetitions of the *first three;* for which reason, I unhesitatingly reject the said *last five* as spurious. This view is strikingly confirmed by MS. P.; for this, the only first-class MS. that is carried on beyond section 40, contains the first three sections on the "umbræ" only. The *second* result is, that if the first three sections on the "umbræ" are to be received, there is good reason

B

why we should consider the possible genuineness of sections 44 and 45 on the "mene mote," which rest very nearly on the same authority.

Now the sections on the "mene mote" have in their favour one strong piece of internal evidence ; for the date 1397 is mentioned in them more than once as being the "root" or epoch from which to reckon. In most cases, the mention of a date 1397 would lead us to attribute the writing in which it occurs to that year or to a *later* year, but a date fixed on for a "root" may very well be a *prospective* one, so that these sections may have been written *before* 1397 ; an idea which is supported by the line "behold wheþer thy date be more or *lasse* þan þe ȝere 1397 ;" sect. 44, l. 5. But I suspect the date to be an error for 1387, since that [see *Somer* in Tyrwhitt's Glossary] was really the "rote" used by Nicholas Lenne. In either case, I think we may connect these sections with the previous sections written in 1391.[1] Besides which, Chaucer so expressly intimates his acquaintance with the subjects of these sections in the Canterbury Tales,[2] that we may the more readily admit them to be really his. There is still less difficulty about admitting the first three sections (41—43) on the "umbræ," because we find similar matter in the treatise of Messahala, from which, as will appear, he derived so much. And hence we may readily conclude that, in the second part, the first forty sections, found in the oldest MSS., are certainly genuine, whilst sections 41—43, as well as 44 and 45, have every claim to be considered genuine also. This need not, however, force us to accept the remaining sections, since they may easily have been added by another hand ; a circumstance which is rendered the more probable by the

---

[1] See Part ii, sect. 1, l, 4 ; sect. 3. l. 11. "Obviously, nobody putting a hypothetical case in that way to a child would go out of his way to name with a past verb [see the second case] a date still in the future."—Morley's Eng. Writers, ii. 282. Similarly, the expression "I wolde knowe," in the former case, precludes a date in the *past* ; and hence we are driven to conclude that the date refers to time present. Curiously enough, there is an exactly parallel case. Blundevill's Description of Blagrave's Astralabe, printed at London by William Stansby, is undated. Turning to his Proposition VI, p. 615, we find— "As for example, I would know the Meridian Altitude of the Sun yᵉ first of July, 1592." The same date, 1592, is again mentioned at pp. 619, 620, 621, 636, and 639, which renders it probable that the book was printed in that year.

[2] " Nother his *collect*, ne his *expans yeres*,
       Nother his *rotes*, ne his other geres ; " l. 11587,8.

fact that sections 41*a*—42*b* merely repeat 41—43 in a more clumsy form, and by the consideration that, if genuine, they should have occupied their proper place immediately after sect. 43, instead of being separated from the former set. As to sect. 46, I pronounce no decided opinion; there is but little to be said either for or against it, and it is of little consequence.

§ 21. But admitting the genuineness of sections 40—45, it at once becomes evident that there are two distinct gaps or breaks in the continuity of the treatise; the first between 40 and 41; and the second between 43 and 44. A little consideration will account for these. Looking at the Canterbury Tales, we observe the very same peculiarity; at certain points there are distinct breaks, and no mending can link the various groups together in a satisfactory manner. This can be accounted for in part by our knowledge of the fact that the poet died before he had completed the proper linking-together of the tales which he had more or less finished; but I think it also shews him to have been a fragmentary worker. It seems very probable that he did sometimes actually tire of a thing which he had nearly completed, and allowed himself to begin something else for which he had meanwhile conceived a newer enthusiasm. Such characters are not uncommon amongst men of great ability. To suppose that, upon reaching " conclusion " 40, he suddenly turned to the sections upon the "umbræ," which are at once more easy to explain, more suitable for a child, and illustrative of a different and more practical use of the Astrolabe, seems to me natural enough; and more probable than to suppose that anything is here lost. For, in fact, it is to the very MSS. that contain sections 41—43 that we are indebted for the last five words of sect. 40, so curiously omitted in the oldest and best MSS.; and this is a direct argument against the supposition of any matter having been here lost.

§ 22. The break between sections 43 and 44 may be explained in a totally different manner. I suppose that the break indicates a *real*, not an accidental, gap. I suppose section 43 to have been really the *last* section of Part ii, and I refer sections 44 and 45 to the *Fourth* Part of the Treatise, and not to the *Second* at all.[1] For if

[1] Not wishing to enforce this view upon every reader, and in order to save

we run through the contents of Parts Three and Four, we observe
that they chiefly involve tables, with reference to one of which we
find the words "vpon wych table ther folwith a *canon*," &c. Now
sections 44 and 45 exactly answer the description ; they are alternative
canons, shewing how certain tables may be used. It happens that "Conclusion" 40 is particularly dependent upon tables. To supply these was
partly the object of Part iv—"the whiche 4 Partie in special shal
shewen a *table of the verray Moeuyng of the Mone from howre to howre,*
euery day and in euery signe, after thin Almenak / *vpon wych table
ther folwith a canon,* suffisant to teche as wel the *maner of the
wyrkyng of þat same conclusion* / as to knowe in owre orizonte
with wych degree of the zodiac that the Mone arisith in any latitude /
& the arising of any planete aftur his latitude fro the Ecliptik lyne."
The opening words of the same Conclusion are—"Knowe be thin
almenak the degree of the Ecliptik of any signe in which þat the
planete is rekned for to be," &c. This is easily said ; but I suppose
that it was not so easy in olden times to know off-hand the exact
position of a planet. It must have been shewn by tables, and these
tables chiefly considered the "mene mote," or average motion of the
planets, and that only for periods of years. If you wanted the
position of a planet at a given hour on a given day, you had to work
it out by figures ; the rule for which working was called a "canon."
This very "canon" is precisely given at length in sect. 44 ; and
sect. 45 is only another way of doing the same thing, or, in other
words, is an alternative canon. When all this is fairly and sufficiently considered, we shall find good grounds for supposing that these
sections on the "mene mote" are perfectly genuine, and that they
belong to Part iv of the Treatise.

I will only add, that the fact of sections 41*a*—42*b* being thus
placed after a portion of Part iv is one more indication that they
are spurious.

§ 23. But it may be objected, as Mr Brae has very fairly
objected, that Conclusion 40 itself ought to belong to Part iv. So it

trouble in reference, I have numbered these sections 44 and 45. But if they
belong, as I suppose, to Part iv, they should have been named "Part iv
Canon 1," and "Part iv, Canon 2" respectively.

ought perhaps, if Chaucer had followed out his own plan. But we have clear indications that his was one of those minds which are not easily bound down to the *exact* completion even of designs which he had himself formed. The Prologue to the Canterbury Tales must have been written later than several of the tales themselves, and yet we find him deliberately proposing to furnish *two* tales for every speaker at a time when he had not even provided for them all once round. The well known difficulty about the number of the pilgrims is probably only one more instance of a similar uncertainty; for the simplest solution of the said difficulty is to suppose that the poet did not exactly know himself, but intended to make it come all right at some vague future period. So in the "Astrolabie," he seems to have laid down a plan, without any very distinct understanding that he was bound to abide by it. It is clear from its contents that the Prologue to the "Astrolabie" was written *before* commencement of the treatise itself, and not, as prefaces generally are, afterwards. He was pleased with his son's progress. Little Lewis had asked him if he might learn something about an astrolabe. The father at once sent him a small astrolabe[1] by way of reward, constructed for the latitude of Oxford, and having 45 circles of latitude on the flat disc (see Fig. 5) instead of having 90 such circles, as the best instruments had.[2] This, however, was a "sufficient" astrolabe for the purpose. But he believes the Latin treatises to be too hard for his son's use, and the conclusions in them to be too numerous. He therefore proposes to select some of the more important conclusions, and to turn them into English with such modifications as would render them easier for a child to understand. He then lays down a table of contents of his proposed five parts, throughout which he employs the future tense, as "the first partie *shal* reherse,"—"the second partie *shal* teche," &c. This use of the future would not alone prove much, but taken in connection with the context, it becomes very suggestive. However, the most significant phrase is in the last line of the Prologue, which speaks of "other noteful

---

[1] "A smal instrument portatif aboute;" Prol. 1. 50.

[2] "The almykanteras in thin Astrelabie ben compowet by two and two." Part ii, sect. 5, l. 1.

thingez, yif god wol vouche sauf & his modur the mayde, mo than I
behete," i. e. other useful things, *more than I now promise, if God and
the Virgin vouchsafe it.* In accordance with his habits of seldom finishing and of deviating from his own plans at pleasure, we have but an
imperfect result, not altogether answerable to the table of contents.
I therefore agree with Mr Brae that the 40th conclusion would have
done better for Part iv, though I do not agree with him in rejecting it
as spurious. This he was led to do by the badness of the text of the
MSS. which he consulted, but we can hardly reject this Conclusion
without rejecting the whole Treatise, as it is found in all the oldest
copies. By way of illustration, I would point out that this is not
the only difficulty, for the Conclusions about astrology ought certainly
to have been reserved for Part v. These are Conclusions 36 and 37,
which concern the "equacions of howses;" and this is probably
why, in two of the MSS. (viz. L. and N.), these two conclusions
are made to come *at the end of the Treatise.* There is nothing for it
but to accept what we have, and be thankful.

§ 24. If, then, the questions be asked, how much of the Treatise
has come down to us, and what was to have been the contents of the
missing portion, the account stands thus.

Of Part i, we have the whole.

Of Part ii, we have nearly all, and probably all that ever was
written, including Conclusions 1—40 on astronomical matters, and
Conclusions 41—43 on the taking of altitudes of terrestrial objects.
Possibly Conclusion 46 is to be added to these ; but Conclusions
41*a*—42*b* are certainly spurious.

Part iii probably consisted entirely of tables, and some at least
of these may very well have been transmitted to little Lewis.
Indeed, they may have been prepared by or copied from Nicholas of
Lynn and John Somer before Chaucer took the rest in hand. The
tables were to have been (and perhaps were) as follows.

1. Tables of latitude and longitudes of the stars which were
represented on the "Rete" of the Astrolabe. Specimens of such
tables are printed in § 30 of this Preface.

2. Tables of declinations of the sun, according to the day of the
year.

3. Tables of longitudes of cities and towns.

4. Tables for setting clocks and finding the meridian altitudes (of the sun, probably).

Such tables as these are by no means lost. There are MSS. which contain little else, as e. g. MS. Hh. 6. 8 in the Cambridge University Library. The longitudes of towns are given in MS. Camb. Ii. 3. 3, at fol. 214b. Again, in MS. F. 25, in St John's College Library, Cambridge, we find tables of fixed stars, tables of latitudes and longitudes of towns, tables of altitudes of the sun at different hours, and many others.

Part iv was to explain the motions of the heavenly bodies, with their causes. This was probably never written. It was also to contain a table to shew the position of the moon, according to an almanac ; and such a table is given in the St John's MS. above mentioned, and in MS. Camb. Ii. 3. 3, at fol. 143. This was to have been followed by a canon, and an explanation of the working of the Conclusion—" to knowe with wych degree of the zodiac that the Mone arisith," and " the arising of any planete," &c. The canon is partly accounted for, as regards the planets at least, by sections 44 and 45, and the " conclusion " by section 40.

Part v was to contain the general rules of astrology, with tables of equations of houses, dignities of planets, and other useful things which God and the Virgin might vouchsafe that the author should accomplish. Sections 36 and 37 tell us something about the equations of houses, but, in all probability, none (or, at least, no more) of this fifth Part was ever written. Tables of equations of houses, for the latitude of Toledo, are given in MS. Camb. Ii. 3. 3, at fol. 177, and elsewhere. Of the general rules of astrology we find in old MSS. somewhat too much, but they are generally in Latin ; however, the Trinity MS. R. 15. 18 has some of them in English.

On the whole, we have quite as much of Chaucer's Treatise as we need care for ; and he may easily have changed his mind about the necessity of writing Part v ; for we actually find him declaring (and it is pleasant to hear him) that "natheles, theise ben obseruauncez of iudicial matiere & rytes of paiens, in which my spirit ne hath no feith ;" ii. 4. 34.

§ 25. I next have to point out the sources whence Chaucer's treatise was derived. Mr Halliwell, in a note at the end of his edition of Mandeville's Travels, speaks of the original treatise on the Astrolabe, written in Sanskrit, on which he supposes Chaucer's treatise to have been founded. Whether the Latin version used by Chaucer was ultimately derived from a Sanskrit copy or not, need not be considered here. The use of the Astrolabe was no doubt well known at an early period in India and among the Persians and Arabs; see the "Description of a Planispheric Astrolabe constructed for Sháh Sultán Husain Safawí, king of Persia," by W. H. Morley, in which elaborate and beautifully-illustrated volume the reader may find sufficient information. Marco Polo says (bk. ii. c. 33) that there were 5000 astrologers and soothsayers in the city of Cambaluc, adding—"they have a kind of *Astrolabe*, on which are inscribed the planetary signs, the hours, and critical points of the whole year;" Marco Polo, ed. Yule, i. 399. Compare also the mention of the instrument in the 161st night of the Arabian Nights' Entertainments, where a translation which I have now before me has the words—"instead of putting water into the basin, he [the barber] took a very handsome astrolabe out of his case, and went very gravely out of my room to the middle of the yard, to take the height of the sun;" on which passage Mr Lane has a note (chap. v, note 57) which Mr Brae quotes at length in his edition. There is also at least one version of a treatise in Greek, entitled περὶ τῆς τοῦ ἀστρολάβου χρήσεως, by Johannes Philoponus, of which the Cambridge University Library possesses two copies, viz. MSS. Dd. 15. 27 and Gg. 2. 33. But it is clear, from his own words, that Chaucer followed the Latin, and I can point out one of the Latin treatises to which he was very considerably indebted. This is the "Compositio et Operatio Astrolabie," by Messahala,[1] of which copies are, I have no doubt, sufficiently numerous. The Cambridge library has four, viz. Hh. 6. 8, Ii. 1. 13, Ii. 3. 3,[2] and Kk. 1. 1, and there

---

[1] Macha-allah or Messahala, an Arabian astronomer, by religion a Jew, flourished towards the end of the eighth century. Latin translations of four of his works (*not* including the Treatise on the Astrolabe) have been printed, and were published at Nuremburg in 1549. A list of his works is given in Casiri (Bibl. Arab. hisp. tom. 1er. pag. 434), and in the Biographie Universelle.

[2] This splendid MS., of the *thirteenth* century, is dated 1276, and illustrated

is another copy in St John's College Library, Cambridge, marked F.
25. The title should be particularly observed; for the treatise is
distinctly divisible into two separate parts, viz. the "Compositio
Astrolabii" and the "Operatio Astrolabii." The former begins with
the words—"Scito quod astrolabium sit nomen Græcum," and explains how to make an astrolabe, and how to inscribe on it the various
necessary lines and circles with sufficient exactness. It is much the
longer portion of the treatise, and (in MS. Ii. 3. 3) is illustrated by
numerous diagrams,[1] whilst the second part has no such illustrations.
But it does not appear that Chaucer made any use of this former
part, as his astrolabe had been procured ready-made. The second
part of the treatise, or "Operatio Astrolabii," begins with the words
"Nomina instrumentorum sunt hec." This is evidently one of the
sources from which Chaucer drew largely, and I have therefore
printed it at length in this volume, from MS. Ii. 3. 3, with a few
corrections from the other copies. Chaucer's Part i is almost wholly
taken from this, but he has expanded it in several places, with the
evident intention of making it more easy to understand. In Part ii,
he has taken from it, with more or less exactness, sections 1—3, 5—
8, 10, 11, 13—18, 20, 21, 24, 25, 27—31, 33—37, 41, and 42;
whilst sections 4, 9, 12, 19, 22, 23, 26, 32, 38—40, and 43
do not appear in it. In other words, Messahala's treatise accounts
for thirty-one conclusions out of forty-three, or about *two-thirds*
of the whole. In some places, Chaucer has translated almost
word for word, so as to leave no doubt as to his authority. Besides
which, I have already remarked that Chaucer's version is *directly*
connected with Messahala by the quotations from the latter which
appear in MS. E.; see description of this MS. above. If it be
inquired, whence did Chaucer derive the remaining third of his
Second Part, I think it very likely that some of it may be found
amongst the varied and voluminous contents of such a MS. as Ii. 3.
3, which is a sort of general compendium of astronomical and
astrological knowledge. The complete solution of this question I
leave to some one with more leisure than myself, being satisfied that

with beautifully-executed coloured diagrams. It is a storehouse of information
about the Astrolabe, and I frequently quote from it.

[1] See the Description of the Plates in this volume.

to have discovered the original of Part i and two-thirds of Part ii is to have made a good start.[1] It must not be omitted that the MSS. of Messahala are *not all alike*, that some copies have propositions which are not in others; and that the order of the conclusions is not invariable. The chief noteworthy difference between Chaucer's version and the Latin original is in the order of the conclusions; it is clear that Chaucer not only took what he liked, but rearranged his materials after his own fashion.

§ 26. About the early printed editions of the Astrolabe, I have not much to say. The Editio Princeps of 1532 was clearly derived from some MS. of the second-class, and, what between the errors of the scribes and printers, absurdities abound. After a careful examination of the old editions, I came to the conclusion that the less I consulted them the better, and have therefore rather avoided them than sought their assistance.

The following is a brief but accurate list of the editions of Chaucer's Works:

1. Ed. by Wm. Thynne, London, 1532. Folio. (The "Astrolabe" begins on leaf ccxcviii, back.)

2. Reprinted, with additional matter, London, 1542. Folio. (Leaf cxxi.)

3. Reprinted, with the matter re-arranged, London, no date, about 1551. Folio.

4. Reprinted, with large additions by John Stowe. London, 1561. Folio.

5. Reprinted, with additions and alterations by Thomas Speght, London, 1598. Folio. (Leaf 261.)

6. Reprinted, with further additions and alterations by Thomas Speght, London, 1602. Folio. (Leaf 249.)

7. Reprinted, with slight additions, London, 1687. Folio. (Page 445.)

8. Reprinted, with additions and great alterations in spelling, &c., by John Urry, London, 1721. Folio. (Page 439.)

Urry's edition is at least as bad as any before it; but there are a

---

[1] The first suggestion as to Chaucer's use of Messahala came to me, as many other excellent suggestions have come to me, from Mr Bradshaw.

few useful explanations in the Glossary, which was added by Mr Timothy Thomas. All these editions not only give the conclusions in a wrong order, but (like the MSS. of the second class) absurdly repeat Conclusion I of Part ii, and reckon the repetition of it as Conclusion III. MSS. of the first class are free from this defect, and may thus be easily known. The only edition worth consulting is that by Mr A. E. Brae, published quite recently, in 1870. Mr Brae made much use of MS. I., besides which he consulted the Printed Editions, and MSS. H. and L. See the descriptions of these MSS. above. From this edition I have taken many hints, and I wish to express, very thankfully, my obligations to it. Mr Brae has brought to bear upon his work much skill and knowledge, and has investigated many points with much patience, minuteness, and critical ability. But I cannot but perceive that he has often expended his labour upon very inferior materials, and has been sometimes misled by the badness of those MSS. to which alone he had access ; whereas I have made a point of consulting MSS. at least half a century older, and far more correct. It is solely for this reason that I believe this edition will be found more generally useful than his, as containing a sounder text ; for I have been so fortunate as to have met with fewer corrupt readings, and in many cases the older MSS. explain passages at once, at the meaning of which he could but guess.[1] It is from no wish to depreciate his labour (which has been considerable), but only for the reader's information, that I point out a few passages where the older MSS. at once correct the text of the Editio Princeps (R) and the printed texts generally.

Conclusion III in R. (which must either be rejected or altered from the form in which it there appears) does not appear at all in the best copies.

Mr Brae observes that the description of the "Moder" (i. 3) is repeated in "all the copies." In the best MSS. it is not so repeated.

The Pin, in R., is said to hold the "tables of the clymathes *in* the

---

[1] For all the information derived from Mr Brae's works, he has my sincere acknowledgments and thanks ; and for any expressions of mine which insufficiently represent his claims as an interpreter of Chaucer, my regret. To all fellow-workers I cordially wish success, and would rather forego all credit than claim too much.

reethe in the wombe of the moder" (i. 14).   But, for the first "in,"
the best MSS. have "and."   The sense is very different.

I here observe, by the way, that, in his Preface, p. 2, Mr Brae
suggests that the Wedge (i. 14) may have been ornamented with the
figure of a horse's head.   This guess is turned into a certainty by the
diagram in MS. Ii. 3. 3, which I have copied.   See Fig. 7.

In the same section (i. 14) we read in R. that this "hors" straineth
all these parts "togyther."   The sense is right enough, but *togyther* is a
mere late gloss.   The best MSS. have the curious Chaucerian phrase
to hepe.   So also in the translation of Boethius, ed. Morris, p. 140—
"god ȝeueþ and departiþ to oþer folk prosperites and aduersites
ymedeled *to hepe ;*" and in Troil. and Cress. iii. 1770 (ed. Tyrwhitt),
we have the complete phrase—"And lost were all, that Love *halt now*
to hepe."   Mr Morris's edition (Aldine Series, iv. 297) has "halt now
to kepe," which is probably a misprint.

In the last part of i. 17, Mr Brae inserts the words *bicause that*
the head of Capricorne, which, he says, are not in the copies.   But
they really *do* exist in the older MSS. ; see i. 17. 34.

In i. 18. 4, where the old MSS. have "is cleped the senyth," Mr
Brae (following R.) prints "is cleped the Signet,"—with the remark
that "Stevins invariably, but very improperly, altered *signet* to
Zenith."   This involves a chronological error of at least three centuries.   Mr Brae occasionally attributes to Stevins or Stöffler expressions which may be found in the Latin version of Messahala,
three hundred years earlier.[1]   It is not a question of opinion, but of
fact.   In this and many similar instances, we must consult the Latin
original, which the reader may now do for himself.

In i. 21, for "the riet of thin astrelabie *with* thy zodiak," R̄.
has "which is thy Zodiake."   The older reading is the better ; for
the Rete is not *identical* with the zodiac, but only *contains* it.

In i. 21. 9, for "by northe the est line," i. e. to the North of the
East line (which is clearly right) R. has "by the north-eest lyne ;"
an obvious corruption of the text.

In i. 21. 42, R. has "transmue" instead of "causen."   But
signs cannot "transmute in us operations."

[1] The double form of the "skale" appears in a MS. dated 1276.

(The curious passage in i. 21. 48—56, found in the old copies, was accidentally omitted in Mr Brae's edition.)

In ii. 3. 29, Mr Brae explains "Alhabor" to be the star Rijel or Rigel (β Orionis). This was because the numbers in the later MSS. are incorrect. But the numbers in the older MSS. are quite consistent with the usual explanation, which identifies Alhabor with Sirius or the Dog-star. That Alhabor and Rigel are totally different appears from the list of stars printed below, from MS. Camb. Ii. 3. 3.[1] As if to preclude all mistake, the diagram in MS. A. represents the Dog-star by a roughly-drawn dog's head, with the name "Alhabor" written on it; see Fig. 2.

In ii. 4. 26, for "infortunyng" R. has "fortune"; this exactly reverses the sense.

In ii. 4. 31, R. omits the necessary words "and þat he be."

In ii. 4. 33, for "ioigned," i. e. joined, R. has "reygned;" which gives no sense.

In ii. 11. 5, R. omits "of any of thise howris þat ben passed, or elles how many howres or partie of."

In ii. 11. 12, for "laste chapitre of the 4 partie," R. has "fourthe partye of the laste chapitre;" the cart before the horse.

In ii. 13. 5, Mr Brae prints "the highest degre," with the note, "in all the copies this word is *lyne*. It ought manifestly to be *degre*." The oldest MSS. have neither *line* nor *degre*, but a third word, viz. *cours*.

So in the rubric to ii. 17, for "longitude" R. has "latitude;" but Mr Brae observes that the object of the problem is *longitude*. The oldest MSS have "longitude" rightly enough.

In ii. 17. 24, R. has "after the syght." Mr Brae well says that "it is difficult to interpret *after the sight*." So it is; but the right reading "after the *site*" is clear enough.

In ii. 23, Mr Brae has an argument to shew that the two stars used were β Ursæ Majoris and the Pole-star; and that the former was the star from which the latitude was derived, whilst the latter (the pole-star) was merely used to help to find the other's place. This

___

See also Fig. 2 and Fig. 9, where they are marked on the lower rim at some distance apart.

curious inversion was caused by the false numbers in the late copies. The true numbers in the early copies shew (as might have been expected) that it was exactly the other way ; the latitude, or rather the elevation of the pole was, naturally enough, derived as usual from observing the pole-star, and the other star (to determine which we have quite insufficient data) was merely used for convenience, to help to fix the pole-star's position.

In ii. 25. 36—40, the old editions are so imperfect that the text has to be guessed at. The old MSS. are clear enough.

In ii. 26. 22, R. has " ouercometh the equinoctial." The right word is "ouerkeruyth," i. e. cuts across, crosses.

In ii. 29. 7, Mr Brae prints " bordure" correctly ; but he had to guess at it, for his authorities had " sonne," which he saw to be absurd.

In ii. 30, he attributes to Stevins the notion that the " wey of the sonne " means the sun's apparent diurnal path, and says that it is wrong. However absurd it may seem, I suspect it is what Messahala means ; at any rate, the oldest MSS. distinctly say " the wey wher as the sonne wente thilke day " (ii. 30. 10) ; but the later copies differ from this.

In ii. 31. 2, the phrase " by north the est " is again corrupted (in R.) into "by Northe-est."

In a note on p. 52, Mr Brae says that Stevins has everywhere wrongly altered *minute* to *Azimuth*. But the latter reading can be defended ; it was so written a century before Stevins was born. The rubric to ii. 34 is corrupt in the later copies ; Mr Brae has restored it by conjecture, and the old copies shew that he has done rightly.

In ii: 34. 6, he has " wayte than of which degre the zodiake is to which the pryck of the altitude of the Mone [applies]." Curiously enough, MS. A. also erroneously has *to which*, but collation at once shews that it is a mere error for *towchith*, and the right reading is as I have given it. R. also wrongly has *to whiche*.

In ii. 35. 18, for " Episicle " R. has " eclyptyke lyne."

In ii. 39. 3, R. has " signet" instead of "lyne Meridional," which cannot well be explained. The last part of ii. 35, viz. ll. 19—27, is very badly represented in R.

The whole of ii. 40 is also so badly represented in R. and the late MSS. that Mr Brae was led to reject it. But it occurs in MSS. A., B., C., and others, and is therefore of the same age as all the Conclusions which precede it.

Besides his print of Chaucer's Astrolabe, Mr Brae has reprinted some curious and interesting critical notes of his own, and has added some essays on Chaucer's " prime," on " the Carrenare," and "shippes opposteres." To some of these I shall refer presently. To all that he has done I am much indebted, and I should, indeed, have abandoned the editing of the present volume but for the fact that I had ascertained the existence of better materials than he happened to meet with.

It is, perhaps, not out of place to observe here that those who are best acquainted with Early English will readily perceive that the spelling, and many turns of expression, are of an older character in the present edition than in any that has preceded it.

§ 27. The works upon, and descriptions of, the astrolabe, are numerous. I have had neither time nor inclination to make researches into the subject ; for which reason I here note the names of a few books which may be examined by the curious reader.

In his Universal Lexicon, Zedler explains that astrolabes are of two kinds, " universal " and " particular." He speaks of the astrolabes (1) of Gemma Frisius ; see Petri Apiani Cosmographia, per Gemmam Phrysium restituta ; (2) of Johan de Rojas, a Spaniard, A. D. 1550; (3) of De la Hire the elder, professor of mathematics at Paris, A. D. 1702 ; (4) of Johannes Stoflerinus (or Stöffler), A. D. 1510. The last of these differed from the others in adopting a different and more convenient system of projection, viz. that upon the plane of the equator, or one parallel to it, the eye being in the antarctic pole, and the arctic pole being made the centre of the instrument. This projection is the same as that which was used by Ptolemy, and it is adopted in the diagrams which accompany Chaucer's treatise in some of the MSS. It should be observed here that the term " astrolabe " alone is vague ; it was originally a general name for any circular instrument used for observation of the stars ; but in the sixteenth and seventeenth centuries it was restricted to the particular kind called the "Astrolabe Planisphere," or astrolabe *on a flat surface*, in which sense alone the word is used throughout this volume. See the English Cyclopædia, Arts and Sciences, s. v. *Astrolabe*.

The simplest work is that by Stöffler or Stoflerinus, as he calls himself; see also Gemma Frisius, Metius, Clavius Bambergensis, the Cursus Mathematicus of Dechales, vol. iv. p. 161, Delambre's History of Astronomy, and other works. The plates in Metius are most exquisitely engraved, and on a large scale, and give a better representation of the instrument than any others that I have seen.

One of the MSS. speaks, as I have said, of an astrolabe belonging to Merton College, Oxford. There is a very nice one, made of brass, and by a Dutch engraver, in the library of King's College, Cambridge. It has several discs or plates, or, as Chaucer calls them, "tables." [1] Of this instrument the same library contains a written description, with some account of the problems it will solve, and an investigation of its probable date, by H. Godfray, Esq., of St John's College. There is also a small silver instrument in Trinity College, Cambridge, which has a circular rim like that of an astrolabe.

There is a book entitled "A verie briefe and most plaine description of Mr Blagrave his Astrolabe," &c., by Mr Blundevill; London, printed by William Stansby. It is undated, but mentions the date 1592 several times. This treatise is very much on Chaucer's plan, as it gives a description of the instrument, followed by the Conclusions which it will solve. But it turns out to be of little practical assistance, because Blagrave's astrolabe was on a different principle. Blundevill, in his Preface, says he has seen but three sorts of astrolabes, first, that of Stofflerus, which was much used for a whole century; secondly, the *Catholicon*, or universal astrolabe of Gemma Frisius; and thirdly, an improved *Catholicon* by Mr Blagrave, "a Gentleman of Reading besides London." He goes on to say that broad astrolabes are bad for use at sea, as being affected by the wind; "which thing to auoyde, the Spaniards doe commonly

---

[1] This word has several senses in Chaucer. It means (1) the discs of an astrolabe; (2) a set of tablets; (3) astronomical tables; and (4) the game of "tables."

make their Astrolabes or Rings narrow and weighty, which for the most part are not much aboue fiue inches broad, and yet doe weigh at the least foure pound." English astrolabes, he says, are very heavy, and six or seven inches broad.   He recommends that more of the southern stars should be represented on the " Rete," such as the Southern Cross, the Southern Triangle, Noah's Dove or Pigeon, and another called Polophilax, lately found out by mariners.   Blagrave's Astrolabe had 71 stars on the Rete, which Blundevill enumerates. He alludes to the division of the mariner's compass into 32 parts, as in Chaucer's time, each part being termed "a Rombe."   He always calls the " rewle " the " Diopter."   There is little else in his volume that illustrates Chaucer.

§ 28. DESCRIPTION OF THE ASTROLABE PLANISPHERE.

There is not, however, much need of reference to books to understand what the astrolabe used by Chaucer was like.   The instrument may be readily understood from a brief description, and from the Plates in this volume.

The most important part of the " astrolabe planisphere " consisted of a somewhat heavy circular plate of metal from four to seven inches in diameter, which could be suspended from the thumb by a ring (i. 1), working with such freedom as would allow the instrument to assume a perfectly perpendicular position (i. 2).   One side of the plate was perfectly flat, and was called the *back*.   This is represented in Fig. 1.   On it was described a number of concentric rings, marked with various divisions, which may be readily understood from the figure. Beginning at the outermost ring, the first two represent the ninety degrees into which each quadrant of a circle can be divided (i. 7). The next two represent the signs of the zodiac, each subdivided into thirty degrees (i. 8).   The next two represent the days of the year, and are rather difficult to mark, as the circle has, for this purpose, to be divided into $365\frac{1}{4}$ equal parts (i. 9). The next three circles·shew the names of the months, the number of days in each, and the small divisions which represent each day, which coincide exactly with those representing the days of the year (i. 10). The two innermost rings shew the saints' days, with their Sunday-letters.   Thus, above the 21st of

C

December, is written "Thome," i. e. St Thomas's day, its Sunday-letter being E; the rest can easily be traced by the tables in Prayer-book (i. 11). These may be thus briefly recapitulated.

1 and 2. Circles of degrees of the quadrant and circle.

3 and 4. Circles of the zodiacal signs, with their degrees.

5 and 6. Circles of the days of the year, with their numbers.

7, 8, and 9. Circles of the months, with their days and numbers of the days.

10 and 11. Circles of saints' days, with their Sunday-letters.

Within all these, are the Scales of Umbra Recta and Umbra Versa, in each of which the scale is divided into twelve equal parts, for the convenience of taking and computing altitudes (i. 12). This primitive and loose method of computation has long been superseded by the methods of trigonometry. Besides these circles, there is a perpendicular line, marking the South and North points, and a horizontal line from East to West.

The other side of the plate, called the *front*, and shewn in Fig. 2, had a thick rim with a wide depression in the middle (i. 3). The rim was marked with three rings or circles, of which the outermost was the Circle of Letters (A to Z) representing the twenty-four hours of the day, and the two innermost the degrees of the quadrants (i. 16). The depressed central portion of the plate was marked only with three circles, the "Tropicus Cancri," the "Æquinoctialis," and the "Tropicus Capricorni" (i. 17); and with the cross-lines from North to South, and from East to West (i. 15). But several thin plates or discs of metal were provided, which were of such a size as exactly to drop into the depression spoken of. The principal one of these, called the "Rete," is shewn in Fig. 2. It consisted of a circular ring marked with the zodiacal signs, subdivided into degrees, with narrow branching limbs both within and without this ring, having smaller branches or tongues terminating in points, each of which denoted the exact position of some well-known star. The names of these stars, as "Alhabor," "Rigel," &c., are (some of them) written on the branches (i. 21). The "Rete" being thus, as it were, a skeleton plate, allows the "Tropicus Cancri," &c., marked upon the body of the instrument, to be partially seen below it. Another form of the "Rete" is shewn in Fig. 9, and other *positions* of the Rete in Fig. 11 and Fig. 12. But it was more usual to interpose between the "Rete" and the body of the instrument (called the "Mother") another thin plate or disc, such as that in Fig. 5, so that portions of this latter plate could be seen beneath the skeleton-form of the "Rete" (i. 17). These plates are called by Chaucer "tables," and sometimes an instrument was provided with several of them, differently marked, for use in places having different latitudes. The one in Fig. 5 is suitable for the latitude of Oxford (nearly). The upper part, above the Horizon Obliquus, is marked with circles of altitude (i. 18), crossed by incomplete arcs of azimuth tending to a common centre, the zenith (i. 19). The lower part of the same plate is marked with arcs denoting the twelve planetary hours (i. 20).

At the *back* of the astrolabe revolved the "rule," made of metal, and fitted with sights, represented in Fig. 3 (i. 13). At the *front* of it revolved the "label," represented in Fig. 6 (i. 22).

All the parts were held together by the central pin (Fig. 4) which passed through the holes in the "moder," plates, "Rete," rule, and label,[1] and was secured by a little wedge (i. 14), which was sometimes fancifully carved to resemble a horse (Fig. 7).

Another "table" or disc is shewn in Fig. 14, and was used for ascertaining the twelve astrological houses.

§ 29. USES OF THE ASTROLABE PLANISPHERE.

I here briefly enumerate such principal uses of the instrument as are mentioned by Chaucer.

The *back* (Fig. 1) shews at once the degree of the zodiac answering to every day in the year (ii. 1). The altitude of the sun can be taken by the "Rule," elevated at the proper angle (ii. 2). If the Rete be properly adjusted to this altitude, we can thus tell the hour of the day (ii. 3). The duration of twilight can be calculated by observing when the sun is 18° below the horizon (ii. 6). Observe the times of sunrise and sundown and the interval is the "artificial day" (ii. 7).

---

[1] "*Pertuis:* m. A hole. *Pertuis de l'Araigne,* the centre of an Astrolabe; the hole wherein all the tables thereof are, by a pin or naile, joined together."—Cotgrave's French Dictionary.

This day, with the duration of morning and evening twilights added to it, is called the "vulgar day" (ii. 9). The plate in Fig. 5 shews the planetary hours (ii. 12). The placing of the sun's degree on the South-line gives the sun's meridian altitude (ii. 13), and conversely (ii. 14). The back of the instrument can shew what days in the year are of equal length (ii. 15). The degree of the zodiac which souths with *any* star can be ascertained by observing two altitudes of the star; but the observations must be made when the star is *very near* the meridian (ii. 17). If the star be marked on the Rete, the said degree is easily found by use of the Rete (ii. 18). We can also find with what degree of the zodiac the same star rises (ii. 19). The use of the Rete also shews the declination of every degree in the zodiac (ii. 20). We can always tell for what latitude a disc such as that in Fig. 5 is constructed, by properly examining it (ii. 21). The latitude of any place can be found by two observations of the altitude of the Pole-star (ii. 23); or of any circum-polar star (ii. 24); or by observing the sun's meridional altitude (ii. 25). The Rete also tells us the "ascensions of signs," or how many degrees of the equinoctial circle pass the meridian with a given sign (ii. 27); as also the "oblique ascensions" of the same (ii. 28). The astrolabe can also be used to discover (but only in an imperfect and approximate manner) the four cardinal points of the compass (ii. 29). We can also compare the altitude of a planet with that of the sun (ii. 30). We can find in what part of the horizon the sun rises (ii. 31); and in what direction to look for a conjunction of the sun and moon (ii. 32); also near what point of the compass the sun is at any given hour (ii. 33). The moon's observed altitude will shew her longitude (ii. 34). We can tell, from two observations of a planet properly made, whether the planet's movement is direct or retrograde (ii. 35). The disc shewn in Fig. 14 helps to shew the "equations of houses" (ii. 36). The four cardinal points can be found *without* an astrolabe, by an experiment properly conducted (ii. 38). The astrolabe can be used to find the degree of the zodiac with which any planet ascends, even when the planet is not situated in the ecliptic (ii. 40).

By the use of the *Umbra Recta* on the back of the instrument, we can take the altitude of an accessible object by a single observation (ii. 41); or of an inaccessible object by two observations (ii. 43). Or, the height of an inaccessible object may likewise be taken by two observations, by the scale marked *Umbra Versa* (ii. 42).

The few conclusions not here referred to are chiefly explanatory, or of minor interest.

§ 30. STARS MARKED ON THE RETE.

Several of the Latin MSS. upon the Astrolabe give a list of the stars marked upon the Rete. The first double list printed below is from the Cambridge MS. which has also furnished us with the Latin version of Messahala. It is given in the form of two tables; the first mentions 49 stars, with the degrees of the zodiac which south along with them, and their declinations from the equinoctial line. The second table mentions some only of *the same* stars, with their longitudes and latitudes, as referred to the ecliptic.

TABLE I. FORTY-NINE STARS MARKED UPON A RETE.

[MS. Camb. Univ. Lib. Ii. 3. 3 ; fol. 70, back.]

Tabula stellarum fixarum que ponuntur in astrolabio, cum gradibus quibus celum mediant, et cum distantia earum ab equinoctiali linea.

Nomina signorum.	Nomina stellarum.	Ymagines stellarum.	Longitudo.		Latitudo.		Pars latitudinis est hec.
			Gr.	Min.	Gr.	Min.	
Aries	(1) Mirach	venter cethi	7*	0	32	30	N †
	(2) Baten kaytoz		18	30	13	30	S
	(3) Panten kai-toz [1]		20	0	14	0	S
	(4) Enif		22	0	23	30	N
	(5) Finis fluxus		29	0	4	30	S

* The MS. has "Gradus 0. Minuta 7; " but I have collated its readings with those in MS. Univ. Lib. Ii. 1. 13, fol. 81, back ; and the latter has "Grad. 7. Min. 0," which seems rather to be meant.

† The MS. has contractions for "Septentrionalis" and "Meridionalis;" I alter these to "N" and "S" throughout, as being more explicit and less troublesome.   [1] Patencataytoz *in* Ii. 1. 13.

Nomina signorum.	Nomina stellarum.	Ymagines stellarum.	Longitudo.		Latitudo.		Pars latitu-dinis est hec.
			Gr.	Min.	Gr.	Min.	
Taurus	(6) Menkar	Naris cethi	6	0	1	0	S
	(7) Algenib	Frons algonis	10	0	49	0	N
	(8) Algecenar[1]		22	0	16	0	S
	(9) Aldebaran	Oculus vel cor tauri	29	0	14	30	N
Gemini	(10) Alhaioth	Hyrcus vel hu-merus sag.	6	0	45	0	N
	(11) Rigil	Pes orionis	11	0	10	0	S
	(12) Algeuze	Humerus dexter orionis	15	0	8	0	N
Cancer	(13) Alhabor	In ore canis merid.	3	0	15	0	S
	(14) Razalgeuze	Cap. d. gemino-rum	9	0	33	0	N
	(15) Algomeyza	In collo canis	13	0	7	0	N
	(16) Markep		21	0	22	30	S
	(17) Egregez		24	0	45	0	N
Leo	(18) Aldurin[2]	In fronte leonis	6	0	6	0	S
	(19) Alfart[3]	Equs vel cingu-lus	13	0	18	30	S
	(20) Calbalezed[4]	Cor leonis	20	0	15	0	N
	(21) Alrucaba	Vrsa	20	0	35	0	N
Virgo	(22) Coruus		1	0	11	0	S
	(23) Dubhe	Id est, Vrsa	2	0	6	0	N
	(24) Deneb alezed	Cauda leonis	15	0	19	30	N
	(25) Algorab	In centauro	22	0	13	30	S
Libra	(26) Alchimec	Inhermis	10	0	7	0	S
	(27) Bennenaz	Filie feretri in themone	9	0	43	0	N
	(28) Alramech	Lanceator	27	0	24	0	N
Scorpius	(29) Alfeta	In corona adri-ane	16	0	29	0	N
	(30) Alachil		17	0	19	0	N
	(31) Yed		26	0	3	0	S
	(32) Calbalacrab	Cor scorpii	27	0	23	0	S
Sagittarius	(33) Alhaue[5]	Capud draconis	13	0	15	0	N
	(34) Rahtaben[6]	Capud serpentis	25	0	51	0	N

[1] Angethanar *in* Ii. 1. 13.   [2] Aldiran *in* MS. Ii. 1. 13.
[3] Alfarth *in* MS. Ii. 1. 13.   [4] Calbelezet *in the same.*
[5] Alhaue vel Razalegue *in the same.*   [6] Razraleyn *in the same.*

Nomina signorum.	Nomina stellarum.	Ymagines stellarum.	Longitudo. Gr.	Min.	Latitudo. Gr.	Min.	Pars latitudinis est hec.
	(35) Wega	Vultur cadens	3	0	38	0	N
	(36) Altair	Vultur volans	16	0	7	0	N
Capricornus	(37) Delfin		29	0	12	30	N
	(38) Alrif	In cigno	29	0	42	0	N
	(39) Addigege	Cauda galline	30	0	43	0	N
	(40) Libedeneb	Cauda capri	6	0	22	0	S
	(41) Delfin	Nubilosior & orientior	10	0	6	0	N
Aquarius	(42) Aldurin [1]		10	0	59	0	N
	(43) Enifelferaz	Musida equi pegasi [2]	13	0	7	0	N
	(44) Denebalgedi	Cauda capricorni	14	0	19	30	S
	(45) Sceath [3]	Crus	30	0	19	0	S
	(46) Alferaz	In pegaso [2]	6	0	24	0	N
Pisces	(47) Mentichel	Humerus equi alati	18	0	25	0	N
	(48) Denebkaitoz	Cauda cethi	22	0	10	0	S
	(49) Sceder [4]		28	0	53	0	N

NOTES. Star (4); Latitude given as 22 in the other MS. (Ii. 1. 13). Star (16); Longitude may be 22. Star (17); Longitude is 34 in the MS. Star (38); Longitude 20 in the other MS. Both the numbers and the directions North and South seem to be occasionally incorrect.

TABLE II.   LONGITUDES AND LATITUDES OF SOME OF THE ABOVE STARS.

Tabula stellarum fixarum uerificatarum per armillos [5] parisius; [6] et est longitudo earum gradus circuli signorum per circulum transeuntem polos zodiaci et stellas ; latitudo vero earum est arcus eiusdem circuli cadens inter stellas et gradus longitudinis earum.

[1] Aldird *in the same.*   [2] *Miswritten* pesagi, pesago.
[3] *Or* (*in other* MS.), Scarath.   [4] *Or*, Seder.
[5] armilla *in other* MS.
[6] The form *parisius*, apparently put for *parisios*, occurs in Barbour's Bruce, iv. 251.

Signa.	Nomina stellarum fixarum.	Ymagines stellarum.	Longitudo.		Latitudo.		Pars latitudinis.
			Gr.	Min.	Gr.	Min.	
Aries	(3) Panta kaytoz	Venter cethi	10	0	20	0	S
	(9) Aldebaran	Oculus vel cor tauri	20	0	5	0	S
Taurus	(7) Algenib	Latus dextrum persei	20	0	30	0	N
	(6) Menkar	Naris cethi	2	0	12	0	S
	(11) Rigil allgeuze	Pes orionis	5	0	30	0	S
Gemini	(10) Alhaioth	Hyrcus	10	0	22	40	N
	(12) Bedelgeuze	Humerus dexter orionis	15	0	15	30	S
	(13) Alhabor	In ore canicule	3	0	39	10	S
Cancer	(15) Algomeiza	In collo canis maioris	14	0	15	30	S
	(14) Razelgeuze	Capud geminorum	8	0	10	0	N
	(23) Dubhe	Vrsa	4	0	50	0	N
Leo	(20) Calbalezed	Cor leonis	18	0	0	10	S
	(19) Alfart	Equs vel singularis cingulus	15	0	22	30	S
Virgo	(24) Denebalezed	Cauda leonis	9	0	12	0	N
	(25) Algorab	Coruus	29	0	15	0	S
	(27) Bennennas	Filie feretri	16	0	53	30	N
Libra	(28) Alramech	Lanceator	13	30	31	30	N
	(26) Alchimech	In-hermis	11	30	20	30	N
Scorpius	(29) Alfeta	In corona	1	20	44	30	N
	(32) Calbalacrab	Cor scorpii	28	0	4	30	S
Sagittarius	(34) Raztaben	Capud draconis	12	0	47	0	S
	(33) Razelgeuze[1]	Capud serpentis	10	0	36	0	N
	(35) Alwega	Vultur cadens	13	0	62	30	N
Capricornus	(36) Altair	Vultur volans	20	0	29	30	S
	(39) Addigege	Cauda galline	21	0	60	30	N
	(44) Denebalgedi	Cauda capri	13	0	2	30	S
Aquarius	(41) Delfin	Nubilosior eius & orientior	6	0	32	0	N
	(43) Enif elferaz	Musida equi pegasi	21	0	23	40	N

[1] Razalegue in other MS

| Signa. | Nomina stellarum. | Ymagines stellarum. | Longitudo. | | Latitudo. | | Pars latitudinis. |
			Gr.	Min.	Gr.	Min.	
	(45) Sceath	Crus [aquarii]	27	0	7	0	S
Pisces	(46) Alferaz mentel	Humerus equi	20	0	31	0	N
	(48) Denebkaytoz	Cauda cethi	21	0	20	0	S

In the above tables I have inserted the numbers (1), (2), &c. for convenience of reference. The 49 stars mentioned are the following. (Compare Ideler, Untersuchungen über die Bedeutung der Sternnamen, &c.) I do not pretend to identify them with perfect exactness.

(1) Mirach; or β Andromedæ.

(2) Perhaps τ Ceti.

(3) ζ Ceti; or, the Whale's Belly; see Ideler.

(4) α Arietis; also called Alnath. *Enif* means *nose*.

(5) A star in Eridanus. But it looks more like ο Ceti.

(6) Menkar; α Ceti; or, the Whale's Nose.

(7) Algenib; or α Persei.

(8) Perhaps γ Eridani.

(9) Aldebaran; α Tauri, or the Bull's Eye.

(10) Capella; α Aurigæ; sometimes called Alhaioth.

(11) Rigel; β Orionis.

(12) α Orionis; often called Betelgeux.

(13) Alhabor; Sirius, or the Dogstar.

(14) Razalgeuze; Castor; α Geminorum.[1]

(15) Algomeisa; Procyon; α Canis Minoris; the Little Dog.

(16) ι Argous; see Ideler. (Different from *Markab*, or α Pegasi.)

(17) Uncertain.

(18) Aldurin; a star in the Lion; uncertain which.

(19) Alphard; Cor Hydræ; α Hydræ.

(20) Calbalased; the Lion's Heart; Regulus; α Leonis.

The name *Razalgeuze* is commonly applied to Pollux; but Castor seems to be meant here; see Ideler, p. 151.

(21) Perhaps a star in Ursa Major; possibly λ Ursæ Majoris, its latitude being wrongly given. *Alrucaba* was also a name for the Polestar (Ideler, p. 14). *Rukhba* means *knee*.

(22) Uncertain; possibly γ Crateræ. See No. 25.

(23) Dubhe; α Ursæ Majoris.

(24) Denebalased; the Lion's Tail; β Leonis.

(25) Algorab; i. e. the Crow; γ Corvi. It is clear that Corvus and Centaurus were not the same as on a modern globe.

(26) Alchimech: Spica Virginis; α Virginis.

(27) Benetnasch; η Ursæ Majoris; the foremost horse in Charles's Wain, which the Arabs likened to a bier with a girl laid on it. (Hence Lat. *feretri filie.*)

(28) Alramech; Arcturus; α Boötis.

(29) Alphecca; α Coronæ Borealis.

(30) Perhaps β Serpentis.

(31) δ Ophiuchi.

(32) Cor Scorpii; the Scorpion's Heart; Antares; α Scorpii.

(33) α Serpentarii vel Ophiuchi; also called Ras Alhagus.

(34) γ Draconis; Etanim; the Dragon's Head.[1]

(35) Wega; Vega; α Lyræ; Lyra.

(36) Altair; α Aquilæ.

(37) Possibly α Delphini; the four stars α, β, γ, δ, are very close together.

(38) A star in Cygnus; very near the next following. In fact, *El-ridf* was only another name for *Arided* (Ideler, p. 74).

(39) Arided; Deneb adigege; α Cygni.

(40) Somewhere near ζ Capricorni.

(41) Probably β Equulei. The name "Delfin" seems to imply that Equuleus was also called Delphinus Minor.

(42) The word "Aldurin" or "Aldira" is meant for Alderamin or α Cephei.

(43) Enif, or Enir; ε Pegasi.

(44) The Goat's Tail; δ Capricorni.

---

[1] It is perfectly clear that the scribe has changed the places of the words "Capud draconis" and "Capud serpentis," or rather, "serpentarii."

(45) δ Aquarii; represented as on the leg of the Waterbearer; hence called "Crus Aquarii."

(46) Alpheraz, or the Horse; β Pegasi.

(47) Mentichel; α Andromedæ. This star and the former are both called "humerus equi."

(48) The Whale's Tail; β Ceti.

(49) Shedir; α Cassiopeiæ.

Even when all allowances are made for the alteration of the position of the pole since this table was made, it must be held to be very faulty. To the numbers given in the "longitude" column we must add always from 7 to 12 degrees, to make them equal to the present longitudes. The second table helps to confirm the interpretation of the first in many cases.

TABLE III. OTHER TABLES OF THE SAME STARS.

After Tables I and II were in type, I found that the tables, as given in MS. Ih. 6. 8, were .very differently arranged, and had the peculiar merit of being *dated*, as well as being explicitly calculated for the latitude of 48½ degrees. ᐧ Their date is A. D. 1223, and I here add them for their curiosity, premising that the extraordinary misspellings of the Arabic names are due to the scribe, and not to me. Thus *Alglari* is for *Algorab; pes canis* means *pes Orionis; Galbaiced* is for *Calbalased; Bacelmara* is an error for *Ras el-marâ*, " the woman's head," α Andromedæ; and *Bacelgohol* is for *Ras Algol* or *Algol*. I may here add, that the word *In-hermis* against star (26) in Table I, is puzzling till explained; nor is it made clearer by being spelt *mermius* (!), as in MS. Ii. 1. 13. It is put for *inermis*, i. e. unarmed, a name given to the hand of Virgo holding the Spica (α Virginis), in Arabic *El-simâkh el-a'zal*, "the unarmed Simâkh;" as distinguished from α Boötis, in Arabic *El-simâkh el-râmih*, "the Simâkh with the lance." See Ideler, p. 51.

[MS. Camb. Univ. Lib. Hh. 6. 8, fol. 236.]

Tabula stellarum fixarum que ponuntur in Astrolabio, certificata ad ciuitatem parisius cuius latitudo est .48. gradus et .30. Minuta. In anno domini nostri iesu christi .1223.

Signa.	Nomina stellarum fixarum.	Gradus cum quibus mediant celum.	Gradus longitudinis ex utraque parte.	Altitudo meridiana.		In quibus ymaginibus sint.
		Gra.	Gra.	Gr.	Min.	
Aries	(3) Pacancaitoz .i. pes caitoz	20	39	28	0	In pede cuiusdam ali- tis (?)
Taurus	(7) Algen	7	71	88	0	In fronte algonis; immo, in dextro persei lateris.
	(9) Aldebaran	28	49	46	30	In oculo tauri.
Gemini	(10) Alhaios	3	74	87	0	In humero agitatoris.
	(11) Ragel	10	39	32	0	In pede orionis.
	(12) Algeuie	14	55	- 49	0	In pede geminorum.
Cancer	(13) Alhaioz	1	36	26	0	In cane meridionali.
	(15) Algomeiza	10	56	48	30	In cane septentrionali.
	(17) Egregez	24	72	87	0	[8. 30.
Leo	(18) Aldiraan	6	52	48	0	In fronte leonis.
	(19) Alfard	13	41	35	0	In ydra serpente .35. 30. ad minus.
	(20) Calbalacet	18	61	57	0	In corde leonis.
	(21) Alrucaba	17	76	90	0	
Virgo	(25) Alglari	18	38	31	0	In centauro.
Libra	(26) Alchimech	9	42	35	0	In-hermis .34.
	(27) Benenah	18	74	84	0	Vltra cenith in temone plaustri.
	(28) Alramech	24	65	65	0	Qui tenet lances.
Scorpius	(29) Elfeta	16	71	72	0	In corona.
	(30) Alielis	17	57	51	0	
	(32) Calbalagrab	27	14	10	0	In corde scorpionis.
Sagittarius	(33) Allahin	13	57	57	0	In capite alay .56. 30. ad plus.
Capricornus	(35) Wega	1	72	79	30	In uulture cadenti.
	(36) Altahir	14	55	48	30	In telo vel in aquila
	(38) Alrif	30	73	84	0	In cigno. [uolanti.
Aquarius	(40) Libideneb	6	34	20	0	In cauda capricorni.
	(41) Delfin	10	45	48	30	
Pisces	(46) Halferaz	6	65	65	0	In pegaso.
	(47) Humerus equi	17	71	87	0	Vltra cenith. In equo alato.
	(48) Dene[b]caitoz	22	36	32	0	In Cauda caytoz.

Tabula stellarum fixarum ; que est longitudo earum a capite arietis, & que latitudo carum ab equatore diei.

Nomina stellarum fixarum maximarum.	Signa.	Longitudo.		Latitudo.		Pars latitudinis.
		Gr.	Min.	Gr.	Min.	
(9) Aldebaran .i. oculus tauri	Taurus	28	2	5	10	S
(11) Raglesiosen .i. pes canis (sic)	Gemini	4	0	31	50	S
(10) Alhaios .i. stella rubea	Gemini	10	23	22	30	S
*Malkanabar .i. scapula canis	Gemini	17	10	17	0	S
(13) Asaare vel Alhabor. hec est stella magna	Cancer	2	40	39	10	S
(15) Algumeiza	Cancer	14	40	16	10	S
(20) Galbaiced .i. cor leonis	Leo	17	40	0	10	N
(24) Neirpha .i. cauda leonis. Magna est	Virgo	9	40	11	50	N
(26) Azimecalazel .i. stella cum lancea	Libra	11	10	2	0	S
(28) Azimecaramech .i. habens lanceam	Libra	12	30	31	30	N
(35) Anazaliaka vel Wega .i. aquila cadens	Capricornus	2	30	62	0	N
*Fonmahout .i. os piscis	Aquarius	22	10	23	1	S
(47) Bacelmara vel rigel .i. caput femine	Aries	2	40	26	0	N
*Bacelgohol .i. capud demonis	Taurus	14	50	23	0	N
(32) Galbaragraph .i. cor scorpionis	Scorpius	27	40	3	0	S
(36) Araranathair .i. aquila uolans	Capricornus	18	30	29	4	N
(39) Panafadigega .i. cauda galline	Aquarius	24	20	9	0	N
(46) Machanastaraz .i. scapula equi	Pisces	17	20	31	0	N
(1) Galbahahot .i. cor piscis, quod quidam uocant genu femine	Aries	9	3	26	20	N

In this list, in which the Arabic words are very badly spelt, as I have said, there are only three stars which do not appear in the other lists. They are marked with an asterisk. The position of the first, Malkanabar, is not clear; the syllable *abar* points to *abûr* and Alhabor, and suggests the star β Canis Majoris; but the position seems nearer to α Leporis. As to the position of the second there can be no doubt, as it is the star Fomalhaut (α Piscis Australis) of the first magnitude. The third is certainly *Algol*, or Medusa's head. The explanation of this may be found in Ideler, p. 88. The Arabs turned Medusa into a demon ; hence the expression " capud demonis."

§ 31. NOTES ON SOME PASSAGES IN CHAUCER.

It is interesting to inquire whether the Treatise on the Astrolabe throws any light upon other passages in Chaucer. This question was taken up by Mr Brae as far back as the year 1851, when he published a series of useful and suggestive articles on the subject in Notes and Queries. Some time afterwards, when making some similar investigations for myself, I came to conclusions of which some were erroneous, and made some mistakes which, if I had sooner become acquainted with Mr Brae's articles, I should not have made.[1] In what I have now to say, I hope the reader will ascribe to Mr Brae's teaching whatever is right, and put down to my own blundering whatever is wrong. I have no desire to claim any credit in the matter, and only make the following observations for the convenience of readers and future editors.

The passages which I quote are cited from the Aldine edition of Chaucer, edited by Dr Morris.

I.   ——" the yonge sonne
        Hath in the Ram his halfe cours ironne."—*Prol.* 7.

The difficulty here really resides in the expression " his halfe cours ; " which means what it says, viz. " his half-course," and not, as Tyrwhitt unfortunately supposed, " half his course." The results of the two

---

[1] I beg leave to assure Mr Brae that the discrepancy which he remarks upon in his edition of the Astrolabe, p. 81, was wholly accidental. I believe it arose from my having read his articles too hurriedly, and missing the point of some of them. I had too much of my own work to do to attend much to the proofs which Mr Furnivall sent me. It was not till some time afterwards that I felt convinced about Mr Brae's explanation of the " Ram."

explanations are quite different. Taking Chaucer's own expression as it stands, he tells us that, a little past the middle of April, "the young sun has run his half-course in the Ram." Turning to Fig. 1, we see that, against the month "Aprilis," there appears in the circle of zodiacal signs, the latter half (roughly speaking) of Aries, and the former half of Taurus. Thus the sun in April runs a half-course in the Ram, and a half-course in the Bull. The former of these was completed, says the poet; which is as much as to say, that *it was past the eleventh of April*.[1]

The sun had, in fact, only just completed his course through the *first* of the twelve signs, as the said course was supposed to begin at the vernal equinox. This is why it may well be called "the *yonge sonne*;" an expression which Chaucer repeats under similar circumstances in the Squyeres Tale, part ii. l. 39.

Chaucer makes the sun enter Aries on the 12th of March (Astrol. ii. 1). In 1865, it entered the sign on the 20th, and in 1871 on the 21st. We thus find a difference of 8 or 9 days between the reckoning in his time and ours. In 1871, the sun entered Taurus on the 20th of April; subtracting 9 days, it entered Taurus, in Chaucer's time, on the 11th of April.[2] This difference is worth remarking.

If the reader wants further confirmation of this view, he may find it in Mr Brae's edition, pp. 65 to 68, and 81 to 84. Compare the expression—"because a sign rises *in the middle of each month*;" p. 47 of Essays on Chaucer, Part I (Chaucer Society).

II.　　"Some wikke aspect or disposicioun
　　　　Of Saturne, by som constellacioun."
　　　　　　　　　　　Knightes Tale, 229.

"But I moste be in prisoun through Saturne;" l. 470.

"My cours, that hath so wyde for to tourne;" l. 1596.

"Myn lokyng is the fadir of pestilens;" l. 1611.

---

[1] This is wholly due to Mr Brae. My own explanation, that Chaucer referred to the *constellation*, not the *sign* of the Ram, I now see to be wrong. Mr Brae shews that Chaucer (and perhaps we may add Lydgate and others) *never* refers to the *constellations*, but *always* to the *signs*. Let this, then, be remembered in future.

[2] This is a sufficiently close approximation for our purpose. The difference between Chaucer's reckoning and ours may be said to vibrate, just at present, between 8 and 9 days. For more exact calculations, the *hour* of the day would have to be taken into account.

Cf. " wykkid planete, as saturne or Mars " ; Astrol. ii. 4. 21 ; notes
in Wright's edition, ll. 2453, 2457 ; and Piers the Plowman, B. vi.
327.　Add to these the description of Saturn—" Significat in . . *quar-
tanis, lepra, scabie,* in mania, *carcere,* . . . *submersione,* &c.　Est infor-
tuna."　Johannis Hispalensis Isagoge in Astrologiam, cap. xv.

> III.　" The thridde night "—*Kn. Ta.* 605.
>> " right as hir day
> Is gerful, right so chaungeth hire aray ; " l. 680.
> "And this day fyfty wykes, fer ne neer ; " l. 992.

See note by me, in Notes and Queries, 4 S. ii. 243, reprinted in Mr
Furnivall's Temporary Preface to Chaucer, p. 103, and Mr Morris's
edition in the Clarendon Press Series, p. 144. , I do not yet see any
point to correct in it. Mr Brae's explanation of " fyfty wykes " must,
however, be also consulted; see Notes and Queries, 1 S. iii. 202, 252. I
make out that the year which would give the days mentioned is 1387.

> IV.　" Allas ! thou felle Mars, allas ! Juno ; " *Kn. Ta.* 701.
> " Nough beth forgeten the *infortune* of Mart ; " l. 1163.
> " By manasyng of Martz, right by figure ; " l. 1177.

Cf. " wykkid planete, as saturne or Mars ; " Astrol. ii. 4. 21 ; " the
infortunyng of an assendent," &c.; ii. 4. 26 ; notes in Wright's edition
to ll. 1749 and 2023 ; and Tyrwhitt's Glossary, s. v. " Puella."　See
also Man of Lawes Tale, 203.

> V.　" As is depeynted in the sterres above; " *Kn. Ta.* 1179.
> " For in the sterres, clerere than is glas," &c. *Man of Lawes Ta.* 96.
> " The heven stood that tyme fortunate."—*March. Ta.* 726.

See Astrolabe, ii. 4 ; cf. Tyrwhitt, note to C. T. 4617.

> VI.　" And after was sche maad the loode-sterre ; " *Kn. Ta.* 1201.
> " Hire sone is eek a sterre, as men may see ; " l. 1203.

Cf. Ovid's Fasti, ii. 153—192 ; especially 189, 190—

> " Signa propinqua micant.　Prior est, quam dicimus Arcton,
> Arctophylax formam terga sequentis habet."

The nymph Callisto was changed into *Arctos*, or the Great Bear.
This was sometimes confused with the other Arctos, or Lesser Bear,
in which was situate the " lodestar " or Pole-star.　Chaucer has followed

this error. Callisto's son, Arcas, was changed into Arctophylax or Boötes; here again, Chaucer says "a sterre" when he means a whole constellation; as, perhaps, he does in other passages. See Smith's Classical Dict. s. v. "Arctos" and "Callisto;" also Ideler's remarks on the Greater and Lesser Bears, in his "Untersuchungen über die Bedeutung der Sternnamen;" pp. xv, and 1—32.

VII. "And in hire hour he walketh forth a paas."—*Kn. Ta.* 1359. See also ll. 1413, 1509.

See Tyrwhitt's note, C. T. 2219; Astrol. ii. 10 and ii. 12. My note on pp. 23 and 24 shews the whole method of working this. Thus, to find the 23rd hour of Sunday, begin with 1, to the left of the upright line, and the 23rd figure is 6, i. e. Venus. Hence, when two hours are still wanting to complete Sunday, we are just beginning the 23rd hour of Sunday, or the hour of Venus. Two hours later we come to sunrise and the figure 2, i. e. Diana; so that Emelye sets off in the first hour of Monday, or the hour of the Moon. Three hours later still, we come upon the figure 3, i. e. Mars, being the fourth hour inequal of Monday, as Tyrwhitt explains.

VIII.   " al his fantasye
            Was torned for to lerne astrologye,
            And cowde *a certeyn of conclusiouns* ;" *Mill. Ta.* 5

         " His almagest, and bookes gret and smale,
            His astrylab[i]e, longyng to his art,
            His augrym-stoones, leyen faire apart ; " *id.* l. 22.

Observe Chaucer's avowal of his disbelief in astrology, Astrol. ii. 4. 36, and Mill. Ta. 265; note the expression "a certein of conclusiouns;" Astrol. Prol. 10; his mention of "ptholome," Astrol. i. 17. 6 (see note to the line), and cf. Wyf of Bath, Prol. 324, and Sompn. Ta. 589; note that the spelling *astrylabe* of the Harl. MS. is clearly wrong; and cf. the expression "nombres in Augrym;" Astrol. i. 9. 3. See also Wright's note to his l. 3210.

IX.   "That now on Monday next, at quarter night,
         Shall falle a reyn ; " *Mill. Ta.* 330.

When all the day of Monday, and a quarter of the night has past, 15 planetary hours are completed, and the 16th is beginning. Now the 16th hour of Monday (see scheme on p. 24) is the hour of *Saturn*.

D

Cf. " Thorwgh *flodes* and þourgh foule wederes · frutes shull faille,
And so sayde *saturne* · and sent ȝow to warne ; "

Piers the Plowman, B. vi. 326.

X. The adjective *rom*, spacious, ample, and its comparative *rommer* (Reeves Tale, 206, 225), occurs again in Astrol. i. 2. 2.

> XI. " Owre hoste sawh [wel] that the brighte sonne
> The arke of his artificial day hath i-ronne,
> The fourthe part, of [*and* ?] half an hour and more, . .
> He wist it was the *eightetene* day
> Of April, that is messanger to May . . .
> And therfore by the schadwe he took his wit
> That Phebus, which that schoon so fair and brighte,
> Degrees was five and fourty clombe on highte,
> And for that day, as in that latitude,
> Hit was *ten* of the clokke, he gan conclude . .
> The fourthe party of this day is goon."

Man of Lawes Prol. 1—17.

For the " artificial day," i. e. the actual duration of the day from sunset to sunrise, see Astrol. ii. 7. The equality of a shadow with its object of course gives an elevation of 45° ; but the reason for alluding to this is made even more clear by noticing that the scale of *Umbra Recta* (Fig. 1) terminates with the equality of the shadow, and with 45°. For *eightetene day*, Tyrwhitt has *eighte and twenty day*, which he could not explain ; see his note. But we must certainly read *eightetene*, as in the Harl. MS. On April 18, the sun was in the 6th of Taurus (see Fig. 1), and the use of a globe [1] will easily shew that the sun's altitude in that degree, at 10 o'clock, was somewhere about 45° or 46° degrees,[2] speaking roughly. But Mr Brae has calculated it exactly, and his results are, that the time when the sun was 45° high on April 18, was 9h. 58m., or only wanting 2 minutes of 10 o'clock. This is even a closer approximation than we might expect, and leaves no doubt as to the correctness of the numbers " *eightetene* " and " *ten*." See Mr Brae's edition of the Astrolabe, pp. 68 and 80.

April 18 in Chaucer's time corresponds to about April 26 now. On April 26, 1871, the sun rose, at London, at 4h. 45m., and set at

---

[1] Any scientific person will naturally object to such a rough way of calculation as resorting to the use of a globe, but I prefer it just *because* it is a rough way of calculating, for we expect no *very* great exactitude *here*. Besides, it is so easy, and so useful in *checking* a closer calculation.

[2] In Mr Furnivall's Temp. Pref. to Chaucer, p. 91, I wrote "about 47 degrees ; " I see now that is too much.

7h. 13m., giving a day of 14h. 28m., the fourth part of which is at
8h. 22m., or, with quite sufficient exactness, at half-past eight. This
would leave a whole hour and a half to signify Chaucer's "half an hour
and more," which, be it observed, was the host's *first* rough guess, *before*
taking a more exact observation. But the matter is made much clearer
by looking at it more closely. How did the host *see* that the 4th part
of the day was past? Of course he looked at the sun. But what did he
know about the sun? He could only (as we shall see) have noted the
point of the horizon at which it rose; for I cannot believe that any one
can do better than accept Mr Brae's equally simple and ingenious explanation, that the host made his guess from observing the extent of
the sun's *azimuthal* arc from sunrise to sunset. The method was incorrect; but we have clear proof (as Mr Brae again rightly suggests) that
Chaucer [1] actually confounded the azimuthal arc with the hour-angle,
in Astrol. ii. 29; see the Additional Note. Set the 6th degree of
Taurus on the E. horizon on a globe, and it is found to be 22° to the
N. of the East point, or 112° from the S. point; doubling this, gives
an azimuthal arc of 224° (exactly as Mr Brae calculates it in his edition,
p. 70); whilst halving it, gives an azimuthal arc from sunrise of 56°.
All, in fact, that the host did, was to observe that the sun had gone
more than half the distance from the point of sunrise to the S. point,
which he might easily do. In numbers, this gives, as was said, an
azimuthal arc of 56° from sunrise, and, therefore, 56° also from the
South. This would happen, as may be seen even by a globe, at about
a quarter past nine; but Mr Brae has made the calculation, and makes
it 20 minutes past nine. This makes Chaucer's "halfe an houre and
more" to stand for half an hour and ten minutes; an extremely neat
result, and confirming the preceding calculations and assumptions. [2]
We conclude then that what "our host saw" was, that the sun had

---

[1] I suppose others did the same. He obviously took it from Messahala,
whom he here follows closely.

[2] It follows that the day mentioned in the opening lines of the Prologue was
either the 17th or the 16th. It was the 17th, if all the tales were told in one
day; it was the 16th, if the Man of Lawe began the second day's series of tales.
I believe Mr Furnivall is right here, and that the Man of Lawe *did* begin the
second day. For how was the host to observe the azimuthal arc of the sun, if
the pilgrims had greatly changed their position since sunrise? And why, if they
had been busy tale-telling, should the host have said, "let us nat mowlen thus

gone more than half-way from his point of rising to the southern point
of the horizon, and he supposed (from his wrong assumption of the
equality of the azimuthal arc with the hour-angle) that more than the
fourth part of the day was gone, by more than half an hour.[1] He
then further observed the sun's altitude to be about 45°, from which
he pronounced it to be ten o'clock. The latter observation was a more
correct and closer one.

> XII. " O firste mevyng cruel firmament,
> With thi diurnal swough that crowdest ay,
> And hurlest al fro est to occident,
> That naturelly wold hold another way."
>
> *Man of Lawes Ta.* 197.

See note in this volume, p. 76 ; and note to the line in " Specimens
of English, A.D. 1298—1393," ed. Morris and Skeat.

> XIII. " Infortunat ascendent tortuous,
> Of which the lordes [*read* lord is] helples falle, alas !
> Out of his angle into the derkest hous ;
> O Mariz Attezere [*read* O Mars, O Atazir], as in this caas ;
> O feeble moone, unhappy been thi paas ;
> Thou knettest the ther thou art nat receyved,
> Ther thou wer wel, fro thennes artow weyved ; "
>
> *Man of Lawes Ta.* 204.

For the word "tortuous," see Astrol. ii. 28. 19 ; the tortuous signs
are from Capricorn to Gemini inclusive ; the most tortuous of these are
Pisces and Aries. Of these two, Aries is the mansion of Mars. We
may then suppose Aries to be the tortuous ascending sign, and the lord
of the ascendent to be Mars ; see Astrol. ii. 4. The "derkest hous"
is perhaps the weakest of the *cadent* houses, or probably the 6th,
which had just set. The "houses," arranged in order of "power,"
are as follows : the four "angles," or the 1st, 10th, 7th, and 4th ; the
"succedents," the 2nd, 11th, 8th, and 5th ; and the "cadents," or
3rd, 12th, 9th, and 6th. In other words, Mars, instead of being in
the ascendent, had "fallen helplessly" beneath the western horizon.
Atazir or *Atacir* is the Spanish spelling of an Arabic word denoting

---

in ydelnesse ? " Perhaps there may be some force too in l. 90—" But of *my* tale
how schal I do *this day ?* " The 16th suits the opening lines even better than
the 17th does. See Note I. above.

[1] The *fact* was, that the fourth part was gone, by at least an hour and a half;
as has been said. But this was a thing which our host could not well have *seen*,
by a mere glance at the sky.

influence; as explained in Dozy, Glossaire des Mots Espagnols dérivés de l'Arabique, p. 207. See note on the line in " Specimens of English, A.D. 1298—1393," ed. Morris and Skeat. With the word " knet-test," cf. " ioigned ; " Astrol. ii. 4. 33. With " receyved," cf. Astrol. ii. 4. 30. With " Ther thou wer wel," cf. " he is wel," Astrol. ii. 4. 34. The exaltation of the Moon was in Taurus ; its depression, or worst position, in the opposite sign of Scorpio. It seems to have been far from its best position.

I subjoin the following extract from Bailey's Dictionary, vol. ii. ed. 1731—" ANGLE (in *Astrology*) certain houses of a scheme of the heavens ; the first house or horoscope is called the angle of the *East ;* the seventh, the angle of the *West ;* the fourth house, the angle of the *North ;* the tenth house, the angle of the *South.*" Bailey is not much to be depended on, so I add another authority.

" DE INVENIENDIS XIII. DOMIBUS.

" In omni hora firmamentum in duodecim partes distribuitur, quæ domus dicuntur, et prima incipit a gradu ascendente, et comprehendit totidem gradus sequentis signi, et sic usque ad duodecim domus per gradus æquales.

" Sed est alia diuisio per inæquales gradus pro terræ latitudine, quæ sic inuenitur. Ponitur gradus ascendentis in oriente, qui est principium primæ domus ; sed septima incipit ab opposito, per gradus totidem ; et gradus tangens lineam meridianam, est initium domus decimæ. Quartum siue imum cœli domicilium incipit ab opposito eius per gradus totidem, *et hæ quatuor domus dicuntur Anguli.* Et gradus inuentus in fine lineæ horæ decimæ, est principium secundæ domus. In fine octauæ inuenitur tertia. In fine quartæ, quinta ; in fine secundæ, sexta ; octaua uero domus per secundum inuenitur, nona per tertiam, et per quintam undecima, per sextam duodecima, per oppositum. Et secunda, quinta, octaua, undecima sunt *post Angulos succedentes.* Tertia, sexta, nona, duodecima sunt *lapsæ, vel cadentes ab Angulis.*"—Epitome Astrologiæ, a Johanne Hispalensi ; cap. xxi.

This useful quotation well illustrates the " Astrolabe," pt. ii. sect. 36, 37 ; it explains the phrase " as in angle," i. e. " as for instance,

in an Angle, or one of the four principal houses;" Astrol. ii. 4. 28;
and also the phrase "in a succedent;" Astrol. ii. 4. 29. Moreover it
suggests that "the meridional angle" is only another name for the
"tenth house;" cf. note XX below.

> XIV.   "Of viage is ther noon eleccioun ...
>   Nought when a roote is of a birthe i-knowe?
>   *Man of Lawes Ta.* 214.

We learn from the third tract in MS. G., p. 10, that there are four
"mobill" signs, Aries, Cancer, Libra, and Capricorn; four "fix"
signs, Taurus, Leo, Scorpio, and Aquarius; and the rest are called
"signes comune." It is added that the right time for going a journey
is when the moon is a "mobill" sign; if it were in a "fix" sign, you
may not go a journey, but you may build a city; if in a "comune"
sign, you may not travel far, nor yet build, but you may safely go to
a city, and live in it; see also Tyrwhitt's note. The whole of Book iv.
of the Epitome Astrologiæ of Johannes Hispalensis is "De Electioni-
bus," and the title of cap. xv. is "Pro itinere." For the word "roote,"
see Astrol. ii. 44, and the Glossarial Index.

> XV.   "Min asce[nde]nt was Taur, and Mars therinne;"
>   *Wyf of Bathes Prol.* 613.

The sign in the ascendent at her birth was Taurus, the mansion of
Venus. Moreover, Taurus was a "feminine" sign. Mars being in
the mansion of Venus is sufficiently significant; see Chaucer's Com-
pleynt of Mars. Cf. "si fuerit [Mars] in Tauro, erit multorum
puerorum," &c. Liber Messahalæ super significationem Planetarum,
cap. iii.

> XVI.   "And thus, god wot, Mercury is desolate
>   In Pisces, wher Venus is exaltate,
>   And Venus faylith wher Mercury is reysed."
>   *Wyf of Bathes Prol.* 703.

The exaltation of Venus is in Pisces, which is also the dejection or
depression of Mercury. The exaltation of Mercury is in Virgo, which
is also the dejection of Venus. This is because the signs Pisces and
Virgo are exactly opposite; see Fig. 1 or Fig. 2. This was explained
by Tyrwhitt in his note on the line.

XVII. " The moone that at noon was thilke day
    That January hadde weddid freissche May
    In tuo of Taure, was into Cancre gliden."—*March. Ta.* 642.

Tyrwhitt altered *tuo* to *ten*, and gave his reason; see his note. He was
wrong in making his calculation from the moon's *mean* motion, as it
differs considerably from her *actual* motion. The question is simply,
can the moon move from the 2nd degree of Taurus to the 1st degree
of Cancer (through very nearly two whole signs, or 59 degrees) in
four days complete (l. 649)? And, in particular, can the moon do this
in the middle of June? Mr Brae (note on p. 93) says decidedly, that
examples of it can be found in every almanack. In one of the volumes
of the Nautical Almanack, I find one very opposite instance, which I
here cite. In June, 1866, the moon's longitude at noon was 30° 22'
on the 9th, and 90° 17' on the 13th; i. e. the moon was in the first
degree of Taurus on the former day, and in the first degree of Cancer
on the latter day, at the same hour; which gives a degree more of
change of longitude than we require. There is therefore no objection
to the reading *tuo*, which the majority of MSS. (I believe) support.

XVIII.   ——" er that dayes eyght
Were passid of the moneth of Juil" (*sic*, wrongly);—*March. Ta.* 888.

  " He [*the sun*] was that tyme in Gemines,[1] as I gesse,
    But litel fro his declinacioun
    Of Canker, Joves exaltacioun."—*id.* l. 978.

The sun had not quite entered Cancer, but was still in Gemini. A
glance at Fig. 1 shews that the sun would enter Cancer about June
12.[2] The former passage must therefore refer to June 8; and the
reading *Juil* is out of the question. We must, of course, read *Juin*,
whatever the scribes of the MSS. may have written to the contrary.
But probably some of the MSS. will be found to have the right
reading.

His " declination of Cancer" means the sun's *maximum* northern
declination, which he attains at the period of the summer solstice,
exactly upon entering Cancer. Now the summer solstice must of
course be in June, not July.

---

[1] Read " Geminis," the ablative plural.
[2] The sun entered Cancer in 1871 on June 21. A difference of 9 days, as
explained above, gives June 12.

Cancer is the exaltation of Jupiter, and the depression of Mars.

The correction *Juin* for *Juil* is due to Mr Brae, and was first published in 1851. See his edition of the Astrolabe, p. 67.

> XIX. "The last Idus of March, after the yeer ;
> Phebus the sonne ful joly was and cleer,
> For he was neigh his exaltacioun,
> In Martes face, and in his mansioun
> In Aries, the colerik, the hote signe ;"—*Squyeres Ta.* i. 39.

"The last Idus" is the very day of the Ides, i. e. March 15. The sun had entered Aries only three days before, on the 12th ; see Astrol. ii. 1. 4. The sun was therefore in the 4th degree of Aries. Aries was called the exaltation of the Sun,[1] and the sun's exaltation was supposed to take place in the 19th degree of the sign in particular, so that he was "nigh his exaltation," and approaching it. The word "face" is technical ; it meant the third of a sign ; see Astrol. ii. 4. 38. In Aries, the first face is that of Mars (where the Sun was), the second that of the Sun, and the third that of Venus. The word *his* in "his mansioun" refers of course, as Tyrwhitt says, to Mars, not to Phebus ; for Aries was the mansion of Mars. The sign Aries is said in MS. G. Tract 3, p. 11, to be *choleric, fiery*, and *masculine ;* cf. Tyrwhitt's note

> XX. "Phebus hath laft [*read* left] the angle merydyonal,
> And yit ascendyng was a best roial,
> The gentil Lyoun, with his Aldryan."—*Sq. Ta.* i. 255.

Four of the astrological houses were called "angles ;" of these, the Southern angle, or "angle meridional," was the tenth house, corresponding to the time from 10 A.M. (at the equinox) to noon. Thus, the sun "leaving the angle meridional" is merely another way of saying that it was past noon. Now, at noon on the 15th of March, in Chaucer's time, the first point of Leo would be on the horizon ; see Mr Brae's edition, p. 87. We need not lay any stress on the word *yit*, which is not always equivalent to the modern *still*, and need not imply any very long continuance.[2] I take the passage to mean merely this, that

---

[1] In Mr Furnivall's Trial-forewords to Chaucer's Minor Poems, there is an unfortunate misprint in footnote 3, p. 87. Read—"Aries is the mansion of Mars, and the exaltation of the Sun," instead of "Venus." The rest of the table is correct.

[2] Only nine lines above, *and yit* is put for *nevertheless ; yet* is used for a very short continuance of time in the Second Nonnes Tale, l. 442, and for a very long period in the Man of Lawes Tale, l. 536.

the sun had passed the meridian, and now the sign Leo, with his Aldryan, was ascending. Considering the frequent shifting of *r* in English, as in *brid* for *bird*, &c., we can have little hesitation in identifying Aldryan with the star *Aldurin* or *Aldiran* mentioned in the "List of Stars marked on a Rete" above; Chaucer makes a much greater change than this, when he turns Ariadne into Adriane.

For determining Aldiran's [1] position, we have, in Table III, its greatest altitude expressed as 48 degrees, for a latitude of about 48 degrees, or a co-latitude of 42 degrees. This makes it 6 degrees north of the equator, showing that in Table I " 6. 0. S " is an error for " 6. 0. N." One table makes its longitude 12 degrees, the other 14 degrees less than that of Cor Leonis. The only star I can see near this position is θ Hydræ, which may have been considered as on the fore-paws of the Lion ; a result which may be illustrated from Lane's Arabic Lexicon, p. 962, which shows that "the Dhirá'án" of the Arabs, or fore-legs of the Lion, were extended even as far as α and β of Gemini, and α and β of Canis Minor. The only remarkable point about this small star is that, near London, it rises but a very little before Cor Leonis, and even at Paris would not long precede it. It may therefore have been looked on as a herald of that celebrated star. But this is mere conjecture, and I leave the working out of this question to others better qualified to do it. The reader should, however, see Mr Brae's remarks in his edition of Chaucer's Astrolabe, pp. 77, 87. If Cor Leonis were on the horizon, the time would be just 2 P.M., which looks as if Chaucer here makes the " angle meridional" to extend for a couple of hours *after* noon instead of *before* it.

It is not clear what authority Speght had for declaring Aldryan to be " a star on the neck of the Lion." In the List already referred to, it is said to be " in fronte Leonis."

---

[1] *Aldiran* is a dual form, and means "the two fore-paws," viz. of the Lion. One of these was called the " extended " paw, and reached as far as α Geminorum ; the other, or " drawn up " paw, was bent so as to end with α and β Canis Minoris. The star named Aldiran must there be sought near the spot whence both fore-legs branch off. Though not very explicit, this does, after a manner, limit its position. The name Aldiran is probably an abbreviation of some longer phrase, just as Algol is put for Ras Algol.

XXI.   "Now dauncen lusty Venus children deere;
       For in the fissch her lady sat ful heyghe,
       And loketh on hem with a frendly eyghe;"—*Squ. Ta.* i. 264.

I take "Venus children" to mean here simply men and women; see
a similar expression in the Knightes Tale, ll. 1628 and 1629. In the
next line, *her* means *their*, corresponding to *hem* for *them* below.
Their lady Venus was then in Pisces; and Pisces was the *exaltation*
of Venus, which explains "full heyghe."   See Tyrwhitt's note to C.
T. 6284.

XXII.   "As rody and bright, as is the yonge sonne
        That in the Ram is *ten* degrees i-ronne."—*Squ. Ta.* ii. 39.

I suppose *ten* is due to some eccentricity on the part of the scribe
of the Harleian MS.   Tyrwhitt has "*foure* degrees;" which can no
doubt be supported by MS. authority. On the day before, March
15, the sun was in the 3rd degree; so that on March 16 he was in
the 4th degree.   "Non heigher was he," in l. 41, means that the sun
was only four degrees above the horizon; cf. ll. 47, 48.

XXIII.   "And this was on the sixte morwe of May . . .
         That yevest, after thy declinacioun,
         To ilk of hem his tyme and his sesoun,
         As that thin herborwe chaungeth low and heighe . . .
         That thou next at this apposicioun [*read* opposicioun]
         Which in the signe shal be of the Leoun," &c.
                                              Frank. Ta. 178—330.

We see from Fig. 1, that the 6th of May is opposite the 23rd
degree of Taurus, which is sufficiently correct.   "After thy declinacioun" means "according to thy declination."   The sun's declination changes from day to day, and with it the solar power and heat;
so that the vegetable kingdom fails or grows according as the sun's
"harbour" or position in the ecliptic makes his daily meridian altitude to be low or high.   The power of the moon over the tides is
referred to in l. 318; and the dependence of lunar upon solar light
in l. 322.   Cf. Astrol. ii. 46.   The highest tides occur when the
sun and moon are either in conjunction or opposition; the latter
is here fixed upon.   But if the sun be in the 23rd degree of
Taurus, the moon, in opposition, is in the 23rd degree of Leo, as
Chaucer says.   If the sun and moon could remain always in opposition, says Aurelius, we might hope to have always a high tide or

"spring-flode" (l. 342). To secure this, the moon must go no faster than the sun (l. 340); and, as the moon in opposition is full, there would be a full moon all the while (l. 341).

> XXIV. "This book spak mochil of operaciouns,[1]
> Touchyng the xxviii. manciouns
> That longen to the mone."—*Frank. Ta.* 401.

The 28 "moon-stations" of the Arabs are given in Ideler's Untersuchungen, p. 287. He gives the Arabic names, the stars that helped to fix their positions, &c. See Mr Brae's edition of the Astrolabe, p. 89. For the influence of the moon in these mansions, we must look elsewhere, viz. in lib. i. cap. xi, and lib. iv. cap. xviii, of the Epitome Astrologiæ of Johannes Hispalensis. Suffice it to say that there are 12 "temperate" mansions, 6 "dry" ones, and 10 "moist" ones.

> XXV. "Phebus wax[2] old, and hewed lyk latoun,
> That in his hoote declinacioun
> Schon as the burned gold, with stremes brighte;
> But now in Capricorn adown he lighte," &c.—*Frank. Ta.* 509.

The sun, in his "hot" or extreme N. declination, shines brightly enough at the summer solstice; but he was now at his lowest altitude, at the winter solstice. He entered Capricorn on the 13th of December, as Chaucer says himself; Astrol. ii. 1. 12. See Fig. 1.

> XXVI. "His tables Tollitanes forth he broughte," &c.—*Frank. Ta.* 537.

See the whole passage.

Here Chaucer mentions the Toletan tables, or tables for the latitude of Toledo; see Tyrwhitt's note. For the "collect" and "expans" years, see Astrol. ii. 44, and the Glossarial Index. Any one who is curious to see such tables may find them in Ptolemy's Almagest, lib. vi. and lib. ix. Ptolemy's expanse years go from 1 to 25, or from 1 to 18, and his collect years by multiples of 25 or 18; whereas Chaucer's go by multiples of 20. For "root," see the Glossarial Index. "Argument" is an astronomical term still in use; see *Entere* in the Glossary. For the "proportionels convenientis,"

---

[1] Printed "of *this* operaciouns," which will not scan. Observe that xxviii. in Old English is always to be read "eight and twenty," never "twenty-eight."

[2] Read "wex" or "wox."

see Astrol. ii. 44. 22.   Next come the lines, which in the Harleian MS. are as follows :—

> "And by his *thre* speeres in his worching,
> He knew ful wel how fer Allnath was schove
> Fro the heed of thilk fix Aries above,
> That in the *fourthe* speere considred is."

There cannot be a moment's doubt that, as Mr Brae well shows in his Preface, p. 13, we must read *eighte* for *thre*, and *ninthe* for *fourthe*.   As the passage stands, it is mere nonsense.   Tyrwhitt has the right readings in both places.[1]   The reader has only to glance at Fig. 10, and he will see at once that the seven inner spheres are spheres of planets.   The eighth is the sphere of fixed stars, and Alnath, being a fixed star (*a* Arietis), was in it.   But the head of the *fixed* Aries, or the true equinoctial point, was in the sphere *above* it, the *ninth* sphere.

The exact amount of the precession of the equinoxes (which is what Chaucer here alludes to) could be ascertained by measuring from time to time the distance between the true equinoctial point and the nearest convenient bright star.   The star Alnath would do well, being of the first magnitude ; indeed, in the time of Hipparchus, its distance from the true equinoctial point was but a few degrees. At the present time, it is " schove " some 35° off, in longitude.   For the word " face " in l. 552, see Note XIX.   Not only every sign, but every "face" had its planet ; hence the phrase " in *whos* face."   A " term " is the Lat. *terminus*.   Besides the division of a sign into three equal parts called *faces*, we find unequal divisions called *terms*. Thus, of Aries, the first 6 degrees are a term of Jupiter, the next 6, a term of Venus ; the next 8, of Mercury ; the next 5, of Mars ; and the last 5, of Saturn.

> XXVII.   " By nature knew he ech ascensioun
> Of equinoxial in thilke toun ;
> For whan degrees fyftene were ascendid,
> Thanne crew he, it mighte not ben amendid ...
> Whan that the moneth in which the world bigan,
> That highte March, whan God first makede[2] man

---

[1] I may just observe that *thre* spoils the scansion of the line, whilst *eight-e* (A.S. *eahta*) is a dissyllable, and suits well enough.

[2] Printed 'makede first,' which scans badly.

> Was complet, and y-passed were also,
> Syn Marche bygan, *tway monthes* and dayes tuo ...
> Cast up his eyghen to the brighte sonne
> That in the signe of Taurus had ironne
> Twenty degrees and oon, and somewhat more ;
> He knew by kynde, and by noon other lore,
> That it was prime, and crew with blisful steven ;
> ' The sonne,' he sayde, ' is clomben up on heven
> *Twenty* degrees and oon, and more i-wis.' "
>
> *Nonne Preestes Tu.* 37,367.

I once proposed an explanation of this which I now entirely give up; it is printed in Mr Morris's small edition of Chaucer for the Clarendon Press, but it is not worth while to repeat it. My difficulty was wholly caused, I now see, by neglecting the word *ech* in the first line. Chaucer says that the cock knew *each* ascension of the equinoxial, and crew at each. That is, he crew every hour, as 15° of the equinoxial make an hour. Chaucer adds that he knew the hour better than an abbey-clock ; see l. 34. This tells us, clearly, that we are to reckon clock-hours, not the unequal hours of the artificial day. Hence the *prime* mentioned below was at a clock-hour, at 6, 7, 8, or 9, suppose. The next point is the date ; and here I am again guided, almost wholly, by Mr Brae's work. The day meant is certainly May 3, because the sun had passed the 21st degree of Taurus ; see Fig. 1. The reading *tuay monthes and dayes tuo* is certainly wrong ; it ought to be *thritty dayes and two*, as in Tyrwhitt, and in Mr Morris's edition for the Clarendon Press. The date, May 3, is playfully denoted by saying that 'March was complete, and also (since March began) thirty-two days more had passed.' The words "since March began" are parenthetical ; and we are, in fact, told that the whole of March, the whole of April, and two days of May were done with. March was then considered the first month in the year, though the year began with the 25th, not with the 1st ; and Chaucer alludes to the idea that the Creation itself took place in March.[1] The day, then, was May 3, with the sun past 21

---

[1] This may be illustrated from the Old English Menologium ; see Grein's Bibliothek der Angelsächsischen Poesie, vol. ii. p. 2.

> " Swylce eác rímcræftige
> On þa ylcan tíd · emniht healdað,

degrees of Taurus. The hour must be had from the sun's altitude,
here said to be '*Twenty* degrees and oon.'   But this is a mere error,
due to the scribe repeating the phrase by mistake ; most MSS. (see
Tyrwhitt's note) have '*Forty* degrees and oon.'   Oddly enough, as
Mr Brae points out, the oldest editions had '*Forty* degrees and oon'
in *both* places, till Francis Thynne corrected the text, and gave the
correct readings. See Thynne's Animaduersions, &c.; ed. G. H. Kings-
ley (E. E. T. S.), p. 50. I again use a globe, and find that the sun would
attain the altitude of 41° nearly at 9 o'clock.   Mr Brae has calculated
it, and makes it 'nine o'clock to the minute.'   It follows that prime,
in this passage, signifies the *end of the first quarter of the day*, reckoned from 6 A.M. to 6 P.M.   What *prime* means in all cases, I do
not pretend to say.   It is a most difficult word, and I think was used
loosely.   It might mean the beginning or the end of a period ; and
the period might be an hour, or a quarter of a day.   I think it was
to obviate ambiguity that the end of the period was sometimes expressed by *high prime,* or '*passid prime,*' or '*prime large ;* ' we also
find such expressions as *half prime, halfway prime,* or *not fully
prime,* which indicate a somewhat long period.   For further remarks,
see Mr Brae's Essay on Chaucer's Prime, in his edition of the Astrolabe, p. 90.   I add some references for the word *prime,* which may
be useful.   We find *prime* in Kn. Ta. 1331 ; Mill. Ta. 368 ; March.
Ta. 613 ; Pard. Ta. 200 ("Long erst than *prime* rong of any belle,"
which goes to show that *prime* was a *fixed* time of day) ; Schip. Ta.
206 ; Sir Thopas, 114 ('fully prime') ; also *passed prime* in Re.
Prol. 52,[1] Fre. Ta. 178, Schip. Ta. 88 ; *prime large* in Sq. Ta. ii. 14.
See also *prime* in Troil. and Cress. ii. 992 (Morris) ; *passed prime,* id.
ii. 1095 ; *an houre after the prime,* id. ii. 1557 ; *prime,* id. v. 15.

XXVIII.   "The sonne fro the south line is descendid
                So lowe, that it nas nought to my sight

Forþan wealdend god · worhte æt frymðe
On þý selfan dæge · sunnan and mónan."
      ' As also arithmeticians
At that very time consider the equinox,
Because all-ruling God wrought at the beginning,
On that very day, the sun and the moon.'
  [1] But Tyrwhitt has *half-way prime,* as in the Six-text edition.

Degrees nyne and twenty as in height.
Foure on the clokke it was, so as I gesse . . .
Therewith the mones exaltacioun
In *mena Libra*, alway gan ascende." . . .

Persones Prol. 2—11.

Besides saying that the sun was 29° high, Chaucer says that his shadow was to his height in the proportion of 11 to 6. Changing this proportion, we can make it that of 12 to $6\frac{6}{11}$; that is, the point of the *Umbra Versa* (which is reckoned by twelfth parts) is $6\frac{6}{11}$ or $6\frac{1}{2}$ nearly. This can be verified by Fig. 1 ; for a straight edge, laid across from the 29th degree above the word "Occidens," and passing through the centre, will cut the scale of Umbra Versa between the 6th and 7th points. The sun's altitude is thus established as 29° above the western horizon, beyond all doubt. Now the day of the month was April 18 (see Note XI) if all the tales were told in one day ; or April 20, if Mr Furnivall's scheme of four days be admitted ; this makes the sun to be either in the 6th or the 9th degree of Taurus. In either case, even the use of a globe will show that the altitude of 29° corresponds closely to four o'clock in the afternoon. Mr Brae gives all the results of his calculations, and makes the altitude of the sun, at 4 P.M. on April 18, 1388 (which is at least near enough, if not the right date altogether), to be 29° 15'. There can therefore be no doubt that the reading *Fowre* is right. Some MSS. have *Ten*, which is out of the question, for that would be *after sunset !* Probably (as Mr Brae suggests) the *tenth hour* may have been meant as a gloss to 'Foure' ; since 4 P.M. is the tenth hour, reckoning from 6 A.M.

We have now to consider the last part of the passage. I make out, merely from the globe, that the point of the zodiac then ascending on the Eastern horizon was about the 4th or 5th degree of Libra. Mr Brae makes the altitude of a certain star which he puts in R. A. 12h. 25m., and N. D. 6° 43', to be 4° 20' ; and I believe my result is much the same as his. At any rate, I feel confident in saying that only some few degrees of Libra had ascended. But, granting all this, how are we to read the passage? Mr Brae proposes to alter it, and to read, " In Libra men al awai gan ascende," which he interprets to mean that the moon was ascending together with the star Min al auwa, which is the 13th of the 28 Arabic 'moon-stations.' The reader should carefully consider all his arguments. I regret that here, and here only, I cannot follow him. The change seems too bold; yet I have nothing better to offer. I merely give my own impression of the matter, which may, after all, be not worth much. *Gan ascende*, in Early English, means no more than *did ascend*, and *alway gan ascende* would mean merely *ever did ascend* or *kept on ascending*. I see nothing unusual in the phrase, though Mr Brae looks upon it as a great difficulty, and objects, in particular, to the word *alway*. Next, Chaucer does not say that the *moon* was ascending, but that the *moon's exaltation* was ascending, which is a very different thing. Again, Chaucer uses *exaltation* in its true astrological sense in other passages (see Notes XVI, XVIII, XIX); but, unfortunately, the Moon's exaltation was in Taurus, a long way off. I have no solution to offer but Tyrwhitt's, that Chaucer did, for once, make a slip (or his scribes have done it for him), and that it ought to be "*Saturnes* exaltation." Next, *In mene libra* would signify *in the middle of Libra*, just as we find "mene mote" (Astrol. ii. 44) for the Latin *medius motus;* but then, the ascending degree was certainly not *in the middle* of Libra, but *near the beginning* of it. This disposes of this reading, and throws us back upon *I mene Libra*, i. e. I refer to Libra; which can be supported by the use of the same phrase—*I mene Venus*—in l. 1358 of the Knightes Tale. This would give—

"Therwith Saturnes exaltacioun,
I mene Libra, alway gan ascende;"

and it would mean no more than that Libra kept on ascending. It had not long before appeared on the horizon. If it be asked, how came Chaucer (or the scribes) to put *mones* for *Saturnes*, I would reply, that a mistake of this kind is easy enough; for the whole of astrology is so technical that no one could be expected to remember it very well; and the *moon* might have been suggested by the fact, that every sign is divided into three equal parts (called *faces*); that the first *face* was then on the horizon; and that the first face of Libra is the face of *the Moon*. Only suppose a momentary confusion between *exaltation* and *face*, and it is done. But this is, perhaps, very unsatisfactory. The reader should also consult Mr Brae's arguments. Perhaps the MSS. may some day help us out here.

XXIX. I add, by way of finishing these notes, the following miscellaneous remarks.

In the Assembly of Foules, l. 59, Chaucer mentions the *nine spheres;* see Fig. 10.

In the Flower and the Leaf (not Chaucer's) we are told that the Sun had just entered the Bull; and that the author rose long before sunrise, at daybreak, about three A.M. This would be about the 11th of April; roughly speaking, daybreak would be about three o'clock, and sunrise about five o'clock.

In the 1st (8th in Morris) stanza of Book ii. of Troilus and Creseide, Phebus is in the Bull on the 3rd of May. The sun would have just completed the 21st degree of Taurus. Cf. Note XXVII.

In the first stanza of the proem to Book iii. of Troilus, there is an address to Venus in "the thridde hevene." Now Venus is in the third heaven, if one begins at the innermost of the nine spheres; for we then have the order, Moon, Mercury, Venus, &c., and, in confirmation of this, we have a mention of the *seventh* sphere, which certainly means that of Saturn; Troil. v. 1823. But it is remarkable that Chaucer *also* adopted the other mode of reckoning, viz. from the seventh sphere of Saturn inwards, giving the order, Saturn, Jupiter, Mars, Sun, Venus, Mercury, Moon. This would place Venus in the *fifth* sphere; and so we find it in stanza ii. of L'Envoy de Chaucer a Scogan. It would also place Mars in the *third* sphere, as in the Compleint of Mars, st. v.[1] In the Boke of the Duchesse, l. 198, is the expression—"a quarter before day." I do not know whether this is a quarter of an hour before day, or a quarter of a day (say three hours) before day. I incline to the latter. Chaucer dreamt about the House of Fame (see l. 111) on the 10th night of December. The winter solstice was then very near at hand.

In the Chanones Yemannes Prologue (l. 272) we have the seven metals belonging to the planets, viz. Saturn, lead; Jupiter, tin; Mars,

---

[1] Not having perceived this *change* in Chaucer's reckoning, I proposed another solution of this expression, which Mr Brae corrected; see Trial Forewords to Chaucer's Minor Poems, by F. J. Furnivall, pp. 85, 121.

E

iron ; Sun, gold; Venus, copper ; Mercury, quicksilver;[1] Moon, silver. Observe that, in the House of Fame, iii. 341, the "Saturnine" Josephus is on a pillar partly of *lead ;* cf. ll. 358, 359. The poets who wrote about warriors are on an *iron* pillar (l. 367), which is the metal of Mars, l. 356. Ovid, the poet of Venus, is on a *copper* pillar, l. 397 ; and so on.

The Prologue to the Legend of Good Women mentions May 1 ; l. 108.

In st. 1 of the Compleint of the Black Knight (which is certainly Lydgate's, not Chaucer's) we have the sun in the middle of the Bull in May. It must mean May 1, when the sun was in the 19th degree of Taurus, nearly.

§ 32. ASTROLOGICAL NOTES.

For a general sketch of Astrology, see the English Cyclopædia, s.v. Worthless as the science is, it is useful to have a few "facts" for handy reference. I therefore attempt a synopsis of the chief points of it, drawn from Johannis Hispalensis Isagoge in Astrologiam.

To save space, I give the information in a tabular form, wherein I denote the 12 Signs by A. T. G. C. L. V. Li. S. Sa. Cp. Aq. P. ; and the Seven Planets Saturn, Jupiter, Mars, Sun, Venus, Mercury, Moon, by St. J. Ms. Sn. V. My. Mo. What the table exactly means shall be explained presently.

Signs.	Man.	Ex.	Day.	Nt.	Com.	Face 1.	Face 2.	Face 3.
A.	Ms.	Sn. (19)	St.	J.	St.	Ms.	Sn.	V.
T.	V.	Mn. (3)	V.	Mn.	Ms.	My.	Mn.	St.
G.	My.	D. H.	St.	My.	J.	J.	Ms.	Sn.
C.	Mn.	J. (15)	V.	Ms.	Mn.	V.	My.	Mn.
L.	Sn.		Sn.	J.	St.	St.	J.	Ms.
V.	My.	My. (15)	V.	Mn.	Ms.	Sn.	V.	My.
Li.	V.	St. (19)	St.	My.	J.	Mn.	St.	J.
S.	Ms.		V.	Ms.	Mn.	Ms.	Sn.	V.
Sa.	J.	D. T.	Sn.	J.	St.	My.	Mn.	St.
Cp.	St.	Ms. (28)	V.	Mn.	Ms.	J.	Ms.	Sn.
Aq.	St.		St.	My.	J.	V.	My.	Mn.
P.	J.	V. (21)	V.	Ms.	Mn.	St.	J.	Ms.

[1] We still have the name *mercury* for quicksilver ; Copper and Venus are both connected with *Cyprus.* Nitrate of *silver* is *lunar* caustic. The sun shines like *gold.* Mars suggested *iron* armour. Saturn's slow motion suggested dull lead.

The first line is to be read thus.

Aries is the mansion (or house) of Mars ; the exaltation (or honour) of the Sun, in the 19th degree of the sign ; the lord of the Triplicity of Aries with its attendant signs is Saturn by day, Jupiter by night, and Saturn in common, both by day and night ; the first Face of Aries (degrees 1 to 10) is that of Mars ; the second Face (degrees 11 to 20) is that of the Sun ; the third Face (degrees 21 to 30) is that of Venus. And so on for the rest ; noting that Gemini is the Exaltation of the Dragon's Head (D. H.), and Sagittarius that of the Dragon's Tail (D. T.).

The meanings of the words are as follows. A *Mansion* or *House* appears to be that sign in which the planet is peculiarly at home for some reason or other.

The *Exaltation* or *Honour* is that degree of a sign in which the planet named has its greatest power ; but the degree was often neglected, and Aries was called the Exaltation of the Sun, simply.

The *Fall* (Lat. *occasus vel detrimentum*) of a planet is the sign opposite its mansion. Libra is opposite Aries ; therefore Libra is the Fall of Mars.

The *Dejection* or *Depression* (Lat. *dedecus*) of a planet is the sign opposite to that of its exaltation. Libra is opposite Aries ; therefore Libra is the Dejection of the Sun. And so on.

A *Triplicity* is a combination of three signs in the form of a triangle, each 120° apart. Thus Aries, Leo, and Sagittarius form the first triplicity ; Taurus, Virgo, Capricorn, the second ; Gemini, Libra, Aquarius, the third ; Cancer, Scorpio, Pisces, the fourth. Equal divisions of a sign (third-parts, namely) are called *Faces*. There were also unequal divisions called *Terms ;* see § 31, Notes XIX, XXVI.

The 'mobill" or movable signs are Aries, Cancer, Libra, Capricorn.

The "fixe" or fixed signs are Taurus, Leo, Scorpio, Aquarius.

The "common" signs are the four others.

The signs Aries, Gemini, Leo, &c. (taking *every other* sign) are *diurnal* or *masculine.*

The rest, Taurus, Cancer, &c., are *nocturnal* or *feminine.*

The first six signs, Aries to Virgo, are *northern* or *sinister* signs.[1]

The last six, Libra to Pisces, are *southern* or *dexter* signs.

The signs Cancer to Sagittarius are *western, sovereign, right,* or *direct* signs. Cf. Astrol. ii. 28, and see Fig. 2.

The rest, Capricorn to Gemini, are *eastern, obedient, tortuous,* or *oblique* signs.

This is all that a reader is likely to want. For other points, see the authorities.

§ 33. DESCRIPTION OF THE PLATES.

Plate I. Fig. 1. The flat back of the Astrolabe ; see Pref. § 28.

Plate II. Fig. 2. The front of the Astrolabe, with raised border. In the wide depression in the middle, the plate called the " Rete " is dropped in, and is shown in its primary position. Other positions of it are sketched in Fig. 11 and Fig. 12.

Plate III. Fig. 3. The " Rewle," carrying two sights, which revolved at the back of the Astrolabe. Astrol. i. 13.

Fig. 4. The central " Pin," shown with the " Wedge " inserted through it. Astrol. i. 14 ; cf. Fig. 7.

Fig. 5. One of the Tables or discs, used by being dropped within the depression on the *front* of the Astrolabe; i. 17. They were marked differently, according to the latitude of the place. The one here drawn is suitable for the latitude of Oxford, nearly.

Fig. 6. The " Label," which revolved at the *front* of the Astrolabe ; i. 22.

Plate IV. Fig. 7. Another form of the " Pin," showing the Wedge cut into the shape of a Horse (i. 14) ; from MS. Camb. Ii. 3. 3.

Fig. 8. Diagram, showing how to draw the three " principal circles ;" see footnote on p. 10.

Fig. 9. Another form of the " Rete," from MS. Ii. 3. 3 ; cf. Fig. 2. This figure shows the " Almury " very clearly ; Astrol. i. 23.

Plate V. Fig. 10. Diagram of the nine spheres; from MS. Camb. Ii. 3. 3. Astrol. i. 17.

Fig. 11. Rough sketch of the position of the " Rete " in Astrol.

---

[1] So called because astrologers looked towards the east or ascendent.

ii. 3 (first part). Denticle opposite C, and first point of Aries opposite X ; 9 A.M.

Fig. 12. Rough sketch of the position of the " Rete" in Astrol. ii. 3 (second part). Denticle near O ; first point of Aries near H ; 8h. 8m. P.M.

Fig. 13. Diagram of the Elevation of the Pole ; Astrol. ii. 23. The arc AN is 56°; A'N is 48°; A'P. is 4°; and PN is 52°. A, A are two positions of the Pole-star.

Plate VI. Fig. 14. A " Table" or disc showing the twelve astrological "Houses;" Astrol. ii. 36 and 37.

Fig. 15. Diagram showing how to ascertain the meridional line from two shadows of an upright gnomon ; Astrol. ii. 38.

Fig. 16. Diagram illustrating the use of the Umbra Recta; Astrol. ii. 41, 41a, and 41b.

Fig. 17. Diagram of the use of the Umbra Versa, at two observations; Astrol. ii. 42, 42a, and 42b.

Fig. 18. Use of the Umbra Recta, at two observations; Astrol. ii. 43, and 43a.

Plate VII. Fig. 19. Diagram showing the influence of the signs upon parts of the human body ; Astrol. i. 21. From MS. Trin. R. 15. 18.

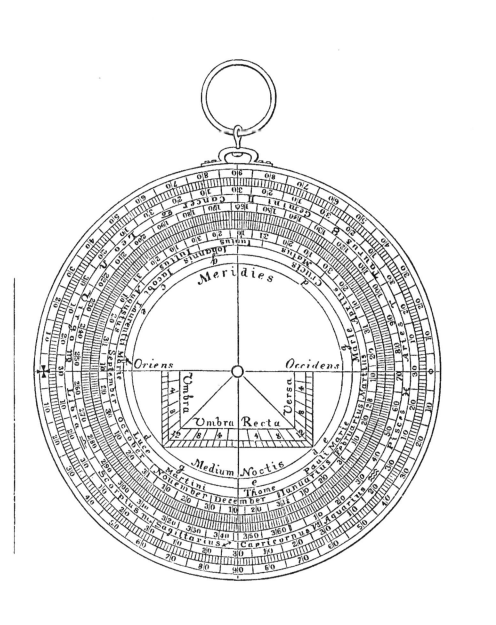

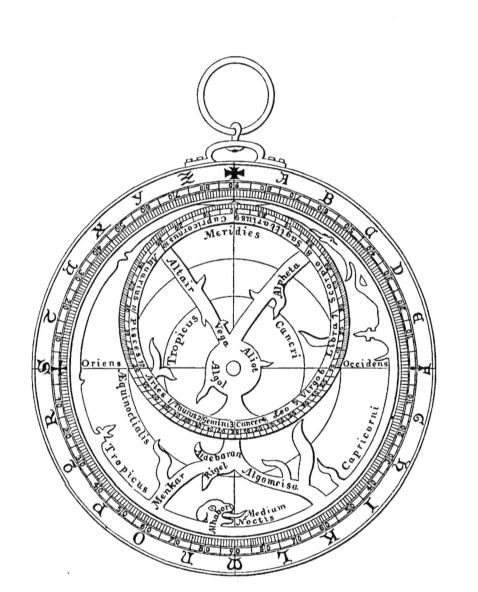

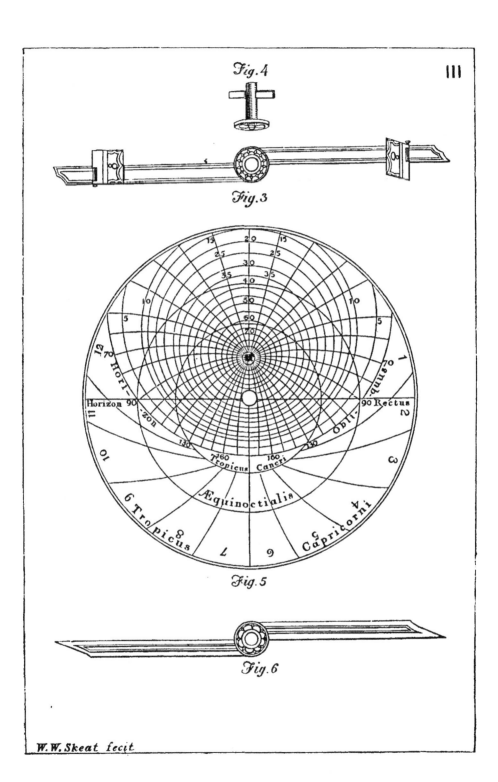

Fig.4

Fig.3

Fig.5

Fig.6

W.W.Skeat fecit

III

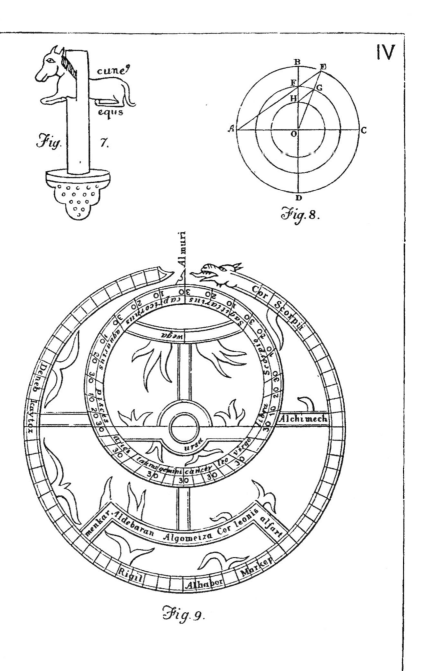

cune⁹

equs

Fig. 7.

B

E

F

G

H

A O C

D

Fig. 8.

Almuri

Cor Scorpij

Deneb kaytoz

sagittarius capricornus aquarius

scorpio pisces

Alchimech

aries taurus gemini cancer leo virgo libra

30 30 30 30 30

vrsa

wega

menkar Aldebaran Algomeiza Cor leonis alfart

Rigil Alhabor Markep

Fig. 9.

W.W. Skeat, fecit

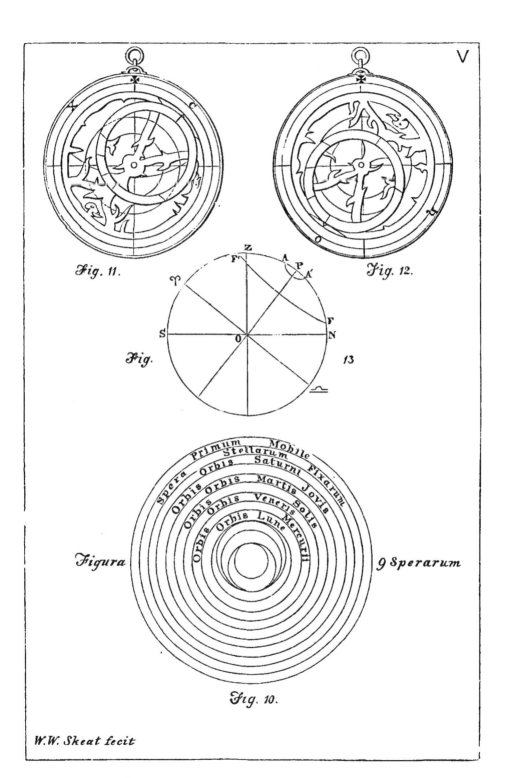

V

Fig. 11.

Fig. 12.

Fig. 13.

Figura   9 Sperarum

Fig. 10.

W.W. Skeat fecit

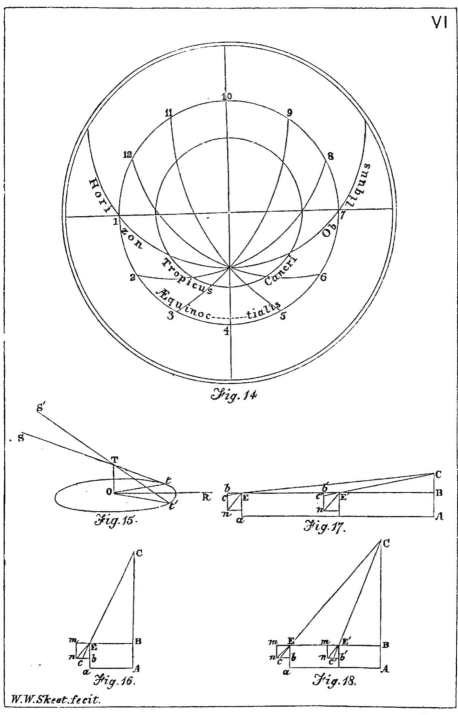

Horizon

Obliquus

Tropicus *Cancri*

Æquinoc------tialis

Fig. 14

S

g'

T

O

R

Fig. 15.

Fig. 17.

Fig. 16.

Fig. 18.

W.W.Skeat.fecit.

VII

W.W.Skeat. del.ᵗ   J.H.Rimbault. Sc.

Fig. 19.

Tractatus de Conclusionibus Astrolabii.

[Bred and mylk for childeren.]

[Fol. 1] Litell Lowys my sone, I haue perceiued well by certeyne
euidences thine abilite to lerne sciencez touchinge noumbres & proporciouns; ¶ & as wel considere I thy bisi preyere in
special to lerne the tretis of the astrelabie. ¶ than, for as mechel as 4
a philosofre seith, ¶ he wrappeth him in his frend, þat condescend-
ith to the rihtful preiers of his frend / ther-for haue I geuen the a
suffisaunt astralabie as for owre orizonte, compowned after the latitude of Oxenford / vp-on which, by mediacion of this litel tretis, I 8
purpose to teche the a certein nombre of conclusions apertenyng to
the same instrument. ¶ I seye a certein of conclusiouns, for thre
causes. ¶ the furste cause is this : ¶ truste wel þat alle the conclusiouns that han ben fownde, or elles possibli myhten be fownde 12
in so noble an instrument as an astralabie, ben vn-knowe perfitly to
any mortal man in this regioun, as I suppose. ¶ a-nother cause is
this ; þat sothly, in any tretis of the astrelabie þat I haue seyn, there
ben some conclusions þat wole nat in alle thinges performen hir by- 16

Little Lewis my son, I perceive that thou wouldst learn the Conclusions of the Astrolabe ; wherefore I have given thee an instrument constructed for the latitude of Oxford, and purpose to teach thee *some* of
these conclusions. I say *some*, for three reasons ; (1) because some of
them are unknown in this land ; (2) because some are uncertain ; or else

hestes ; ¶ & some of hem ben to harde to thy tendre age of .x. yer
to conseyue. ¶ this tretis, diuided in 5 parties, wole I shewe the
vnder ful lihte rewles & naked wordes in englissh ; for latyn ne
20 kanstow yit but smal, my lite sone. ¶ but natheles, suffise to the
thise trewe conclusiou*n*s in englissh, as wel as suffisith to thise noble
clerkes grekes thise same conclusiou*n*s in grek, ¶ & to arabiens in
arabik, ¶ & to Iewes in Ebrew, & to the latyn folk in latyn / whiche
24 latyn folk han hem furst owt of othre diu*er*se langages, & writen in
hir owne tonge, þat is to sein, in latyn. ¶ & god wot, þat in alle
this[e] langages, & in many mo, han thise conclusiou*n*s ben suffisantly lerned & tawht / & yit by diu*er*se rewles, ryht as diu*er*se
28 pathes leden diu*er*se folk the rihte wey to Roome. ¶ Now wol I
prey mekly eu*er*y discret p*er*sone þat redith or herith this litel tretis,
to haue my rewde endytyng for excused, & my sup*er*fluite of wordes,
for two causes. ¶ the firste cause is, for that curio[u]s enditing &
32 hard sentence Is ful heuy atones for swich a child to lerne. ¶ &
the seconde cause is this, þat sothly me semeth betre to writen vn-to
a child twies a good sentence, than he for-get it ones. ¶ And lowis,
ȝif so be þat I shewe the in my lihte Englissh as trewe conclusiou*n*s
36 touching this mat*er*e, & nawht only as trewe but as many & as subtil
[* Fol. 1 b.] conclusiou*n*s as ben shewed in latyn *in ani co*m*mune tretis
of the astrelabie / kon me the more thank ; ¶ and p*r*eye god saue the
kyng, þat is lord of this langage, & alle that him feyth bereth & obeieth,
40 eu*er*ech in his degree, the more and the lasse. ¶ but co*n*sidere wel, that
I ne vsurpe nat to haue fownde this werk of my labour or of myn
engin. ¶ I nam but a lewd co*m*pilatour of the labour of olde Astrolog[i]ens, and haue hit translated in myn englissh only for thi doc-
44 trine ; ¶ & with this swerd shal I slen envie.

---

(3) are too hard. This treatise, divided into five parts, I write for thee
in English, just as Greeks, Arabians, Jews, and Romans were accustomed
to write such things in their own tongue. I pray all to excuse my
shortcomings ; and thou, Lewis, shouldst thank me if I teach thee as
much in English as most common treatises can do in Latin. I have
done no more than compile from old writers on the subject, and I have
translated it into English solely for thine instruction ; and with this
sword shall I slay envy.

¶ The firste partie of this tretis shal reherse the figures & the membres of thin Astrolabie, by-cause þat thow shalt han the grettre knowyng of thin owne instrument.

¶ The second partie shal teche the werken the verrey practik of 48 the forseide conclusiouns, as ferforth & as narwe as may be shewyd in so smal an instrument portatif a-boute. ¶ For wel wot euery astrologien þat smalest fraccions ne wol nat ben shewid in so smal an instrument, as in subtil tables calkuled for a kawse.   52

¶ The .3. partie shal contienen diuerse tables of longitudes & latitudes of sterres fixe for the Astrolabie, ¶ & tables of declinacions of the [sonne], & tables of longitudes of Citeez & of townes ; ¶ & as wel for the gouernance of a clokke as for to fynde the altitude Meridian / 56 & many [a]-nother notable conclusioun, aftur the kalendres of the reuerent clerk[e]s, frere I. Somer & frere N. Lenne.

¶ The .4. partie shal ben a theorik to declare the Moeuynge of the celestial bodies with [þe] causes. ¶ the whiche 4 partie in special 60 shal shewen a table of the verray Moeuyng of the Mone from howre to howre, euery day *& in euery signe, after thin Almenak / [* Fol. 2.] vp-on wych table ther folwith a canon, suffisant to teche as wel the maner of the wyrkyng of þat same conclusioun / as to knowe in 64 owre orizonte with wych degree of the zodiac that the Mone arisith in any latitude / & the arising of any planete aftur his latitude fro the Ecliptik lyne.

¶ The .5. partie shal ben an introductorie aftur the statutz of owre 68 doctours, in which thow maist lerne a gret part of the general rewles of theorik in Astrologie.   ¶ in which .5. partie shaltow fynde tables of equacions of howses aftur þe latitude of Oxenford ; ¶ & tables of dignetes of planetes / & other noteful thingez / yif god wol vouche 72 sauf & his modur the mayde, mo than I be-hete, &c.

The *first* part gives a description of the instrument itself.

The *second* teaches the practical working of it.

The *third* shall contain tables of latitudes and longitudes of fixed stars, declinations of the sun, and the longitudes of certain towns.

The *fourth* shall shew the motions of the heavenly bodies, and especially of the moon.

The *fifth* shall teach a great part of the general rules of astronomical theory.

𝕳𝖊𝖗 𝖇𝖞𝔤𝖞𝖓𝖓𝖊𝖙𝖍 𝖙𝖍𝖊 𝖉𝖊𝖘𝖈𝖗𝖎𝖕𝖈𝖎𝖔𝖓 𝖔𝖋 𝖙𝖍𝖊 𝕬𝖘𝖙𝖗𝖊𝖑𝖆𝖇𝖎𝖊.

1. ¶ Thyn Astrelabie hath a ring to putten on the t[h]owmbe of thy ryht hand in takyng the heyhte of thynges. ¶ & tak kep, for from hennes-forthward, I wol clepe the heyhte of any thing þat is taken by thy rewle, the altitude, wit*h*-owte mo wordes.

2. ¶ This ring rennyth in A Maner turet, fast to the Moder of thyn Astrelabie, in so Rowm a space þat hit desturbith nat the in-strument to hangen aft*ur* his rihte centre.

[Fol. 2 b.] **3.** ¶ The moder of thin Astrelabie is [þe] thikkeste plate, p*er*ced wit*h* a large hole, þat resseyuyth in hir wombe the thynne plates *com*powned for diu*er*se clymatz, & thi Riet shapen in manere of a net or of a webbe of a loppe ; & for the more declaraci*oun*, lo here the figu*re*.

4. ¶ This Moder is deuyded on the bakhalf wit*h* a lyne, þat cometh dessendinge fro the ryng down to the nethereste bordure. ¶ the whiche lyne, fro þe for-seide Ryng vn-to the centre of the large hole amydde, is cleped the sowth lyne, or elles the lyne M*e*ridional. ¶ & the remenant of this lyne downe to the bordure is cleped the north lyne, or elles the lyne of Midnyht.   ¶ & for the more declara-ciou*n*, lo here the figu*re*.

Here begins the *first* part ; i. e. the description of the Astrolabe itself.

1. *The Ring.* See figs. 1 and 2.   The Latin name is *Armilla sus-pensoria ;* the Arabic name is spelt *alhahuacia* in MS. Camb. Univ. Ii. 3. 3, but Stöffler says it is *Alanthica, Alphantia,* or *Abalhantica.*   For the meaning of " rewle," see § 13.

2. *The Turet.* This answeit nearly to what we call an *eye* or a *swivel.* The metal plate, or loop, to which it is fastened, or in which it turns, is called in Latin *Ansa* or *Armilla Reflexa,* in Arabic *Alhabor.*

3. *The Moder.* In Latin, *Mater* or *Rotula.* This forms the body of the instrument, the back of which is shewn in fig. 1, the front in fig. 2. The " large hole " is the wide depression sunk in the front of it, into which the various discs are dropped.   In the figure, the " Rete " is shewn fitted into it.

4. See fig. 1 ; Chaucer describes the "bak-half" of the instrument first.   The centre of the " large hole amydde " is the centre of the instru-ment, where a smaller hole is pierced completely through.   The *Sowth lyne* (marked *Meridies* in figs. 1 and 2) is also called *Linea Meridiei ;* the *North lyne* is also named *Linea Mediæ Noctis.*

[Fol. 3] 5. ¶ Ouer-thwart this for-seide longe lyne, ther crosseth hym a-nother lyne of the same lengthe from est to west. Of the whiche lyne, from a lityl croys + in the bordure vn-to the centre of the large hole, is cleped the Est lyne, or elles the lyne Orientale ; ¶ & the 4 remenant of this lyne fro the forseide + vn-to the bordure, is cleped the west lyne, or the lyne occidentale. ¶ now hastow her the 4 quarters of thin astrolabie, deuyded after the 4 principals plages or quarters of the firmament. ¶ & for the more declaracioun, lo here 8 thi figure.

6. ¶ The est side of thin Astrelabie is cleped the riht side, ¶ & the west side is cleped the left side. ¶ for-get nat this, lite[l] lowys. ¶ put the ring of thin Astralabie vp-on the thowmbe of thy ryht hand, and thanne wole his right side be toward thy left side, & his 4 left side wol be toward thy right side ; tak this rewle general, as wel on the bak as on the wombe-side. ¶ vp-on the ende of this est lyne, as I first seide, is marked a litel + ; wher as euere-mo generaly is considered the entring of the first degree in wich the sonne arisith. 8 ¶ & for the more declaracioun, lo here þe figure.

[Fol. 3 b.] 7. ¶ Fro this litel + vp to the ende of the lyne Meridional, vndur the ryng, ¶ shaltow fynden the bordure deuyded with 90 degres ; & by that same proporcioun is euery quarter of thin Astrolabie deuyded. ¶ ouer the wiche degrees ther ben nowmbres of 4 augrym, þat deuyden thilke same degres fro 5 to 5, as shewith by longe strykes by-twene. ¶ of wyche longe strykes the space by-twene contienith a Mile-wey. ¶ & euery degree of the bordure contieneth 4 Minutes, that [is] to seyn, minutes of an howre. ¶ & 8 for more declaracioun, lo here the figure

---

5. The *Est lyne* is marked with the word *Oriens;* the *West lyne,* with *Occidens.*

6. The rule is the same as in heraldry, the *right* or *dexter* side being towards the spectator's left.

7. As the 360 degrees answer to 24 hours of time, 15° answer to an hour, and 5° to twenty minutes, or a *Mile-way,* as it is the average time for walking a mile. So also 1° answers to 4 minutes of time. See the two outermost circles in fig. 1, and the divisions of the "border" in fig. 2.

[Fol. 4] **8.** ¶ Vnder the *com*pas of thilke degres ben writen the names of the 12 signes, as Aries, taurus, gemini, Cancer, leo, virgo, libra, Scorpio, Sagittarius, Ca*pri*cornus, Aquarius, Pisces ; & the 4 nombres of the degres of tho signes ben writen in Augrim aboue, & w*ith* longe deuysiou*n*s, fro 5 to 5 ; deuyded fro tyme þat the signe entreth vn-to the laste ende. ¶ but vnderstond wel, þat thise degrees of signes ben eue*r*ich of hem considered of 60 Mynutes, & eue*r*y 8 Minute of 60 secondes, & so forth in-to smale fraccions infinit, as seith Alkabuci*us*. ¶ & ther-for, know wel, þat a degree of the bordure contienith 4 Minutis, and a degre of a signe contienith 60 Mynutis, & haue this in Mynde. ¶ & for the more declaraciou*n*, lo here thi 12 figu*re*.

9. ¶ Next this folwyth the cercle of the dayes, þat ben figured in maner of degrees, þat contienen in-nowmbre 365 ; dyuyded also w*ith* longe strikes fro 5 to 5, & the nombre[s] in Augrym writen 4 vnder þat cercle. ¶ and for more declaraciou*n*, loo heere thy figure.

[Fol. 4 b.] **10.** ¶ Next the cercle of the dayes folweth the Cercle of the names of the Monthes ; þat is to seyen, Ianuar*e*, Februar*e*, Marcius, April*e*, Mayus, Iuyn, Iulius, Augustus, Septembre, Octo- 4 ber, Novembr*e*, Decembr*e*. ¶ the names of thise Monthes wer*e* cleped in Arabyens, so*m*me for hir p*ro*pretes, & some by statutz of lordes, some by other lordes of Rome. ¶ ek of thise Monthes, as liked to Iulius cesar & to cesar Augustus, some were *com*powned of diuerse 8 nombres of dayes, as Iuyl and August. ¶ thanne hath Ianuar*e* xxxi daies, Februar*e* 28, March 31, Aprill*e* 30, May 31, Iuni*us* 30, Iulius 31, Augustus 31, September 30, Octobr*e* 31, Nouembr*e* 30, Decem- ber 31. ¶ natheles, al-thow that Iulius cesar tok 2 daies owt of 12 Feu*e*rer & put hem in his monith of Iuylle, & August*us* cesar cleped

---

8. See the third and fourth circles (reckoning inwards) in fig. 1.
9. See the fifth and sixth circles in fig. 1.
10. See the seventh, eighth, and ninth circles in fig. 1. The names of the months are all Roman. The month formerly called *Quinctilis* was first called *Julius* in B.C. 44 ; that called *Sextilis* was named *Augustus* in B.C. 27. It is a mistake to say that Julius and Augustus made the alterations spoken of in the text ; what Julius Cæsar really did, was to add 2 days to the months of January, August (Sextilis), and December,

the Monyth of August After his name & ordeyned it of 31 daies, yit truste wel, þat the sonne dwelleth ther-for neuere the more ne lesse in on signe than in another.

11. ¶ Than folwen the names of the halidayes in the kalender & next hem the lettres of the A. b. c. on wich they fallen. ¶ & for the more declaraciou*n*, loo here thi figure.

[Fol. 5]   12. ¶ Next the forseide cercle of the A. b. c., vnder the cros-lyne, is Marked the skale, in Man*er* of 2 Squyres or elles in Man*ere* of laddres, þat scruith by hise 12 poyntes & hise deuisiou*n*s of ful many a subtil conclusiou*n*. Of this forseide skale, fro the 4 croos-lyne vn-to the verre angle, is clepid *vmbra* [*versa*], & the nether partie is cleped the *vmbra* [*recta*, or elles *vmbra extensa*.] ¶ & for the more declaraciou*n*, loo here the figure.

13. ¶ Thanne hastow a brod Rewle, þat hath on either ende a Square plate p*er*ced w*ith* a *c*ertein holes, some more & some lesse, to resseyuen the stremes of the sonne by day, and ek by mediaciou*n* of

---

and 1 day to April, June, September, and November. February never had more than 28 days till he introduced bissextile years.

11. See the two inmost circles in fig. 1. The names given are adopted from a comparison of the figures in the Cambridge University and Trinity MSS., neither of which are quite correct. The letters of the " A. b. c " are what we now call the Sunday letters. The festivals marked are those of St Paul (Jan. 25), The Purification (Feb. 2), The Annunciation (Mar. 25), The Invention of the Holy Cross (May 3), St John the Baptist (June 24), St James (July 25), St Lawrence (Aug. 10), The Nativity of the Blessed Virgin (Sept. 8), St Luke (Oct. 18), St Martin of Tours (Nov. 11), and St Thomas (Dec. 21).

12. The " skale " is in Latin *Quadrans*, or *Scala Altimetra*. It is certain that Chaucer has here made a slip, which cannot be fairly laid to the charge of the scribes, as the MSS. agree in transposing *versa* and *recta*. The side-parts of the scale are called *Umbra versa*, the lower part *Umbra recta* or *extensa*. This will appear more clearly at the end of Part II.

13. See fig. 3, Plate III. Each plate turns on a hinge, just like the "sights" of a gun. One is drawn flat down, the other partly elevated. Each plate (*tabella vel pinnula*) has two holes, the smaller one being the lower. This *Rewle* is named in Arabic *Alhidada* or *Allidatha;* in Latin *Verticulum*, from its turning easily on the centre ; in Greek *Dioptra*, as carrying the sights. The straight edge, passing through the centre, is called the *Linea Fiduciæ*. It is pierced by a hole in the centre, of the same size as that in the *Mother*.

4 thyn Eye, to knowe the altitude of sterres by nyhtc. ¶ & for the more declaracioun, lo here thi figure.

14. ¶ Thanne is ther a large Pyn in maner of an Extre, þat goth thorow the hole / that halt the tables of the clymates & the Riet in the wombe of the Moder / thorw wich pyn ther goth a litel wegge 4 which þat is cleped the hors, þat streynet[h] alle thise parties to hepe; this forseide grete Pyn in maner of an extre is ymagyn[e]d to be the Pol Artyk in thin Astralabie. ¶ & for the more declaracioun, lo here the figure.

[Fol. 5 b.] **15.** ¶ The wombe-side of thyne Astrelabie is also de-uyded wit[h] a longe croys in 4 quarters from est to West, fro sowth to north, fro riht side to left side, as is the bak-side. & for the more 4 declaracioun, lo here thi figure.

16. ¶ The bordure of which wombe-side is deuyded fro the poynt of the est lyne vn-to the poynt of the south lyne vnder the ring, in 90 degres; & by þat same proporcioun is euery quarter de-4 uyded as ys the bak-side, þat amounteth 360 degres. ¶ & vnderstond wel, þat degres of this Bordure ben answering & consentrik to the degrees of the Equynoxial, þat ys deuyded in the same nombre as euery othere cercle is in the heie heuene. ¶ This same bordure is 8 deuyded also with 23 lettres capitals & a smal croys + aboue the south lyne, þat shewith the 24 howres equals of the clokke / &, as I haue said, 5 of thise degrees maken a Milewey, & 3 Milewey maken an howre. ¶ & euery degree of this bordure contineth 4 Mynutes, &

---

14. See fig. 4, Plate III. The *Pin* is also called *Axis* or *Clavus*, in Arabic *Alchitot;* it occupies the position of the Arctic or North Pole, passing through the centre of the plates that are required to turn round it. The *Wedge* is called *cuneus*, or *equus restringens*, in Arabic *Alpheraz* or the horse, because it was sometimes cut into the shape of a horse, as shewn in fig. 7, Plate IV, which is copied from MS. Univ. Camb. Ii. 3. 3.

15. See fig. 2, Plate II. In the figure, the cross-lines are partly hidden by the *Rete*, which is separate and removable, and revolves with-in the border.

16. The *Border* was also called *Margilabrum, Margolabrum,* or *Limbus.* It is marked (as explained) with hour-letters and degrees. Each degree contains 4 minutes *of time,* and each of these minutes contains 60 seconds *of time.*

euery Minut 60 secoundes ; now haue [y] told the twye. ¶ & for the 12
more declaracioun, lo here the figure.

[Fol. 6]   17. ¶ The plate vnder thi Riet is descriued with 3 [prin-
cipal] cerclis ; of wiche the leste is cleped the cercle of cancer / by-
cause þat the heued of cancer turnyth euermor consentrik vp-on the
same cercle. ¶ in this heued of cancer is the grettest declinacioun 4
northward of the sonne. ¶ & ther-for is he cleped the Solsticioun
of somer ; ¶ wiche declinacioun, aftur ptholome, is 23 degrees & 50
Minutis, as wel in cancer as in Capricorne. this signe of cancre is
cleped the tropik of Somer, of *tropos*, þat is to seyn Agaynward, for 8
thanne by-gynneth the sonne to passe fro vs-ward ; ¶ & for the
more declaracioun, lo here the figure.

[Fol. 6 b.] ¶ The Middel cercle in wydnesse, of thise 3, is cleped
the cercle equinoxial / vp-on whiche turneth euermo the hedes of 12
aries & libra. ¶ & vnderstond wel, þat euermo this cercle equinoxial turnyth Iustly fro verrey est to verrey west ; ¶ as I haue
shewed the in the sper solide. ¶ this same cercle is cleped also the
weyere, *equator*, of the day / for whan the sonne is in the heuedes of 16
aries & libra, than ben the daies & the nyht[es] illike of lenghthe in
al the world. ¶ & ther-fore ben thise two signes called the equinoxiis. ¶ & alle þat Moeuyth with-in the heuedes of thise aries &
libra, his Moeuyng is cleped north-ward / & alle that Moeuyth with- 20
oute thise heuedes, his Moeuyng is clepid sowth-ward as fro the
equinoxial. ¶ tak kep of thise latitudes north and sowth, & for-get
it nat. ¶ by this cercle equinoxial ben considered the 24 howres of
the clokke ; for eueremo the arisyng of 15 degrees of the equinoxial 24
maketh an howre equal of the clokke. ¶ this equinoxial is cleped

---

17. We may place under the *Rete* any plates we please. If only the
Mother be under it, without any plate, we may suppose the *Mother*
marked as in fig. 2. The plate or disc (*tympanum*) which was usually
dropped in under the *Rete* is that shewn in fig. 5, Plate III, and which
Chaucer now describes. Any number of these, marked differently for
different latitudes, could be provided for the Astrolabe. The greatest
declination of the sun measures the obliquity of the ecliptic, the true
value of which is slightly variable, but was about 23° 31′ in Chaucer's
time, and about 23° 40′ in the time of Ptolemy, who certainly assigns to
it too large a value. The value of it must be known before the three

the gyrdelle of the firste Moeuyng, or elles of the *angulus primi motus vel primi mobilis*. ¶ And *nota*, þat firste Moeuyng is cleped "Moeu-
28 yng" of the firste Moeuable of the 8 spere, whiche Moeuyng is fro est to west, ¶ & eft agayn in-to est / also it is clepid "girdel" of the first Moeuyng, for it departeth the furste Moeuable, þat is to seyn, the spere, in 2 ilike parties, euene distantz fro the poles of this
32 world.

¶ The wydeste of thise 3 principal cerkles is cleped the cercle of capricorne, by-cause þat the heued of capricorne turnyth euermo con-sentryk vp-on the same cercle / in the heued of this for-seide capri-
36 corne is the grettest declinacioun sowthward of the sonne, & ther-for is it cleped the solsticioun of wyntur. this signe of capricorne is also cleped the tropik of wyntur, for thanne bygynnyth the sonne to come agayn to vs-ward. ¶ & for the more declaracioun, lo here thi·
40 figure.

[Fol. 7] **18.** ¶ Vp-on this forseide plate ben compassed certein cerclis þat hihten Almicanteras / of which som of hem semen perfit cercles, & somme semen inperfit. the centre þat standith a-Middes the
4 narwest cercle is cleped the senyth ; ¶ & the netherest cercle, or the furste cercle, is clepid the orisonte, ¶ þat is to seyn, the cercle

---

circles can be drawn. The method of finding their relative magnitudes is very simple. Let ABCD (fig. 8, Pl. IV) be the tropic of Capricorn, BO the South line, OC the West line. Make the angle EOB equal to the obliquity (say 23½°), and join EA, meeting BO in F. Then OF is the radius of the Equatorial circle, and if GH be drawn parallel to EF, OH is the radius of the Tropic of Cancer. In the phrase *angulus primi motus*, *angulus* must be taken to mean angular motion. The "first moving" (*primus motus*) has its name of "moving" (*motus*) from its denoting the motion of the *primum mobile* or "first moveable." This *primum mobile* (by some placed in the *ninth* sphere) is here identified with the *eighth* sphere, or *sphæra stellarum fixarum*. See the fig. in MS. Camb. Univ. Ii. 3. 3 (copied in fig. 10, Plate V). Some authors make 12 heavens, viz. those of the 7 planets, the *firmamentum* (*stellarum fixarum*) the *nonum cælum, decimum cælum, primum mobile*, and *cælum empyræum*.

18. See fig. 5, Pl. III. This is made upon the alt-azimuth system, and the plates are marked according to the latitude. The circles, called in Latin *circuli progressionum*, in Arabic *Almicantarath*, are circles of al-titude, the largest imperfect one representing the horizon (*horizon obli-quus*), and the central dot being the zenith, or pole of the horizon. In

þat deuydeth the two Emysperies / þat is, the partie of the heuene
a-boue the Erthe & the partie be-nethe.   ¶ thise almy-kanteras ben
compowned by 2 & 2, al be it so þat on diuers Astrelabies some   8
almy-kanteras ben deuyded by on / & some by two, & somme by .3.
aftur the quantite of the Astrelabie.   ¶ this forseide cenyth is ymagened to ben the verrey point ouer the crowne of thyn heued / &
also this senyth is the verrey pool of the orisonte in euery regioun. 12
¶ & for the more declaracioun, lo here thi figure.

19. ¶ From this senyth, as it semeth, ther come a maner
krokede strikes like to the clawes of a loppe, or elles like to the werk
of a womancs calle, in keruyng ouerthwart the Almykanteras.   ¶ &
thise same strikes or diuisiouns ben cleped Azymuthz.   ¶ & 4
they deuyden the Orisonte of thin astrelabie in 24 deuisiouns.   ¶ &
thise Aʒimutz seruen to knowe the costes of the firmament   ¶ & to
othre conclusiouns, as for to knowe the cenyth of the sonne & of
euery sterre.   ¶ & for [the] more declaracioun, lo here thi figure.   8

[Fol. 7 b.] **20.** ¶ Next thise azymutz, vnder the cercle of Cancer,
ben ther 12 deuysiouns embelif, moche like to the shap of the
azimutes, þat shewen the spaces of the howres of planetes / & for
mor declaracioun, lo here thi figure.   4·

21. ¶ The riet of thin Astrelabie with thy zodiak, shapen in
maner of a net or of a lop-webbe aftur the olde descripcioun, which
thow maist tornen vp and down as thi-self likyth, contienith certein
nombre of sterres fixes, with hir longitudes & latitudes determynat ; 4

---

my figure, they are "compounded by" 5 and 5, but Chaucer's shewed
every second degree, i. e. it possessed 45 such circles.   For the method
of drawing them, see Stöffler, leaf 5, back.

19. Some Astrolabes shew 36 of these azimuthal circles, as in my
figure. See Stöffler, leaf 13, where will be found also the rules for drawing them.

20. If accurately drawn, these *embelife* or oblique lines should divide the portions of the three circles below the *horizon obliquus* into
twelve equal parts.   Thus each arc is determined by having to pass
through three known points.   They are called *arcus horarum inequalium*,
as they shew the "howres inequales."

21. In fig. 2, Pl. II, the *Rete* is shewn as it appears when dropped
into the depression in the front of the instrument.   The shape of it
varied much, and another drawing of one (copied from Camb. Univ. MS.

¶ ȝif so be þat the makere haue nat erred.   the names of the sterres
ben writen in the Margyn of the riet ther as they sitte ; ¶ Of whiche
sterres the smale poynt is cleped the Centre / And vnderstond also
8 þat alle sterres sittinge wyth-in the zodiak of thin astrolabie ben
cleped sterres of the north, ¶ For thei Arisen by northe the est lyne.
¶ & alle the remenant fixed, owt of the zodiak, ben cleped sterres of
the sowth ; ¶ but y sey nat þat they arisen alle by southe the est
12 lyne ; witnesse on aldeberan & Algomeysa.   ¶ generally vnderstond
this rewle, that thilke sterres þat ben cleped sterres of the north
arisen rather than the degree of hire longitude, & alle [the] sterres of
the sowth arisen aftur the degree of hire longitude ; this is to seyn,
16 sterres fixed in thin Astralabie. the mesure of this longitude of
sterres is taken in the lyne Eclyptik of heuene, vnder which lyne,
whan that the sonne & the Mone ben lyne-riht or elles in the
superfice of this lyne / than is the Eclips of the sonne or of the
20 [* Fol. 8] Mone ; as y shal declare, & ek the cause why. but *sothly the
Ecliptik lyne of thy zodiak is the owttereste bordure of thy zodiak,
ther the degrees ben marked.

¶ Thy zodiak of thin Astralabie is shapen as a compas wich þat
24 contienith a large brede, as aftur the quantite of thin astralabie / in
ensample þat the zodiak in heuene is ymagened to ben a superfice
contienyng a latitude of 12 degrees, ¶ wher[as] al the remenant of
cerkles in the heuene ben ymagined verrey lynes with-owte eny
28 latitude.   ¶ Amiddes this celestial zodiak ys ymagined a lyne, which
þat is cleped the Ecliptik lyne / vndur which lyne is euermo the
wey of the sonne.   ¶ Thus ben ther 6 degrees of the zodiak on þat
on side of the lyne, and 6 degrees on that other.   ¶ This zodiak is

---

Ii. 3. 3, fol. 66 *b*.) is given in fig. 9, Pl. IV.   The positions of the stars
are marked by the extreme points of the metal tongues.   Fig. 2 is taken
from the figures in the Cambridge MSS., but the positions of the stars
have been corrected by the list of latitudes and longitudes given by Stöffler, whom I have followed, not because he is *correct*, but because he
probably represents their positions as they were supposed to be in
Chaucer's time very nearly indeed.   There was not room to inscribe the
names of all the stars on the *Rete*, and to have written them *on the plate
below* would have conveyed a false impression.   A list of the stars marked
in fig. 2 is given at the end of the volume.   The Ecliptic is the circle

deuided in 12 principal deuisiouns, þat departen the 12 signes. ¶ &, 32
for the streitnes of thin astrelabie, than is euery smal deuysioun in a
signe departid by two degrees & two ; I Mene degrees contenyng 60
Minutes.   ¶ & this forseide heuenissh zodiak is cleped the cercle of
the signes / or the cercle of the bestes / for *zodia* in langage of grek 36
sownyth ʻbestesʼ in latyn tonge.   ¶ & in the zodiak ben the 12
signes þat han names of bestes ; or elles for whan the sonne entrith
in any of the signes, he taketh the proprete of swich bestes ; or
elles for þat the sterres that ben there fixed ben disposed in signes of 40
bestes, ¶ or shape like bestes ; ¶ or elles whan the planetes ben vnder
thilke signes, þei causen vs by hir influence operaciouns & effectes lik
to the operaciouns of bestes.   ¶ & vnderstonde Also, þat whan an
hot planete comyth in-to an hot signe, than encresseth his hete ; & 44
ȝif a planete be cold; thanne amenuseth his coldnesse, by-cause of the
hote signe.   ¶ & by this *conclusioun* maistow take ensample in alle
the signes, be they moist or drie, or moeble or fix ; rekenyng the
qualite of the planete as I furst seide. ¶ & euerich of thise 12 Signes 48
hath respecte to a certein parcelle of the body of a man and hath it
in gouernance ; as aries hath thin heued, & taurus thy nekke & thy
throte / gemyni thyn armholes &. thin armes, ¶ & so forth ; as
shal be shewed more pleyn in the 5 partie of this tretis. this zodiak, 52
which þat is part of the 8 spere, ouer-kerueth the equinoxial ; and
he ouer-kerueth hym again in euene parties / & þat on half declinith
sowthward, & þat other northward, as pleynli declareth the tretis of
the spere.   ¶ & for mor declaracioun, lo here thi figure.   56

[Fol. 8 b.] 22. ¶ Thanne hastow a label, þat is schapen lik a rewle,

---

which crosses the Equinoctial at its East and West points.   In Chaucer's
description of the zodiac, carefully note the distinction between the Zodiac of the Astrolabe and the Zodiac of Heaven.   The former is only *six*
degrees broad, and shews only the northern half of the heavenly zodiac,
the breadth of which is *imagined* to be 12 degrees.   Chaucer's zodiac
only shewed *every other* degree in the divisions round its border.   This
border is divided by help of a table of right ascensions of the various
degrees of the ecliptic, which is by no means easily done. See *Additional
Note* on this section ; which explains Fig. 19, Plate VII.   I may add
that the *Rete* is also called *Aranea* or *Volvellum;* in Arabic, *Alhancabuth.*
   22. *The Label.* See fig. 6, Pl. III.   The *label* is more usually used

saue þat it is streit & hath no plates on either ende w*ith* holes;
¶ but w*ith* the smale point of the forseide label, shaltow kalcule
4 thyne equaciou*n*s in the bordure of thin Astrolabie as bi thin almury.
¶ & for the more declaraciou*n*, lo here thy fig*ur*e.

23. ¶ Thin almury is cleped the denticle of cap*r*icorne or elles
the kalkuler.   ¶ this same Almury sit fix in the hed of capricorne, & it seruyth of many a necessarie conclusiou*n* in equaciou*n*s
4 of thynges, as shal be shewed; ¶ & for the more declaraciou*n*, lo here
thi fig*ur*e.

Her endith the descripcion of the Astrelabie.

[Fol. 9] Her bygynnen the conclusions of the Astrelabie.

1. To fynde the degree in which the sonne is day by day, after hir cours a-bowte.

[Hic incipiunt conclusiones astrolabii; & prima est ad inveniendum gradus solis in quibus singulis diebus secundum cursum
sol est existens.]

¶ Rekene and knowe which is the day of thi monthe ¶ & ley
thi reule vp that same day, & thanne wol the v*er*ray point of thy
rewle sitten in the bordure, vp-on the degree of thy sonne.   ¶ En-
4 sample as thus; the yer of oure lord 1391, the 12 day of March at
Midday, I wolde knowe the degree of the sonne.   ¶ I sowhte in the
bakhalf of myn astrelabie, and fond the sercle of the daies, ¶ the
which I knowe by the names of the Monthes writen vndur the same
8 Cercle.   ¶ Tho leide I my rewle ou*er* this forseide day, & fond the

-----

on the *front* of the instrument, where the *Rete* and other plates revolve.
The *rule* is used on the *back*, for taking altitudes by help of the scale.

23. *The Almury;* called also *denticulus, ostensor,* or "calculer." In
fig. 2, it may be seen that the edge of the *Rete* is cut away near the head
of Capricorn, leaving only a small pointed projecting tongue, which is
the almury or denticle, or (as we should now say) pointer. As the *Rete*
revolves, it points to the different degrees of the border. See fig. 9.

Part II, § 1. [The Latin headings to the propositions are taken from
the MS. in St John's College, Cambridge.] See fig. 1. Any straight

point of my rewle in the bordure vp-on the firste degree of aries, A litel with-in the degree / & thus knowe I this conclusioun. ¶ A-nother day, I wolde knowe the degree of my sonne, & this was at Midday in the 13. day of decembre; I fond the day of the monthe in maner 12 as I seide / tho leide I my rewle vp-on this forseide 13. day, & found the point of my rewle in the bordure vp-on the first degree of capricorne / a lite with-in the degree ¶ & than haddy of this conclusioun the ful experience / & for the more declaracioun, lo her thi 16 figure.

[Fol. 9 b.] ## 2. To knowe the altitude of the sonne, or of othre celestial bodies.

[De altitudine solis & aliorum corporum supra celestium.]

¶ Put the ring of thin Astrelabie vp-on thi riht thowmbe & turne thi lift side a-gayn the light of the sonne/ And rem[e]ue thi rewle vp and down til þat the stremes of the sonne shyne thorgh bothe holes of thi rewle. ¶ loke thanne how Many degrees thi 4 rewle is a-reised fro the litel crois vp-on thin est line, & tak ther the altitude of thi sonne. ¶ & in this same wyse maistow knowe by nyhte the altitude of the Mone, or of brihte sterres / this chapitre is so general euer in on, þat ther nedith no more declaracion; but 8 for-get it nat. ¶ & for the more declaracioun, lo here the figure.

[Fol. 10] ## 3. To knowe euery tyme of the day by liht of the sonne, & euery tyme of the nyht by the sterres fixe, & eke to knowe by nyht or by day the degree of any signe þat assendith on the est Orisonte, which þat is cleped communly the assendent or elles oruscupum.

---

edge laid across from the centre will shew this at once. Chaucer, reckoning by the old style, differs from us by about 8 days. The first degree of Aries, which in his time answered to the 12th of March, now vibrates between the 20th and 21st of that month. This difference of eight days must be carefully borne in mind in calculating Chaucer's dates.

2. Here " thy left side " means the left side of thine own body, and therefore the right or Eastern edge of the Astrolabe. In taking the altitude of the sun, the rays are allowed to shine through the holes; but the stars are observed by looking through them. See figs. 1 and 3.

[Ad cognoscendum quidlibet tempus diei per solis indicacionem,
& quodlibet tempus noctis per quasdam stellas in celo fixas;
ac eciam ad inveniendum & cognoscendum signum super orizontem qui communiter vocatur ascendens.]

¶ Tak the altitude of the sonne whan the list, as I haue said;
¶ & set the degree of the sonne, in kas þat it be by-forn the Middel
of the day, among thin al-my-kanteras on the est side of thin
4 astralabie; & ȝif it be after the Middel of the day, set the degree of
thy sonne vp-on the west side / tak this manere of settyng for a
general rewle, ones for euere. ¶ & whan thow hast set the degree
of thy sonne vp as many Almykanteras of heyhte as was the alti-
8 tude of the sonne takyn by thi rewle, ¶ ley ouer thi label, vp-on the
degree of the sonne; ¶ & thanne wol the point of thi label sit[t]en in
the bordure, vp-on the verrey tid of the day. Ensample as thus /
the yer of owre lord 1391, the 12 day of March, I wold knowe the
12 tyd of the day. I tok the altitude of my sonne, ¶ & fond þat it
was 25 degrees and 30 of Minutes of heyhte in the bordure on the
bak-side. ¶ tho turnede I Myn astrelabie, & by cause that it was
by-forñ Midday, I turnede Mi riet and sette the degree of the sonne,
16 that is to seyn the .1. degree of Aries / on the riht side of myn
Astralabie, vp-on þat 25 degrees & 30 of Minutes of heyhte among
myn almy-kanteras / tho leide I my label vp-on the degree of my
[* Fol. 10 b.] sonne, & fond the poynte *of my label in the bordure, vp-on
20 a capital lettre þat is cleped an X; tho rekened I alle the capitalles
lettres fro the lyne of Midnyght vn-to this forseide lettre X, & fond
þat it was 9 of the clokke of the day. tho loked I do[w]n vp-on the
Est Orisonte, and fond there the 20 degree of gemynis assending;

---

3. Drop the disc (fig. 5) within the border of the mother, and the *Rete*
over it. Take the altitude by § 2, and let it be 25½°. As the altitude
was taken by the *back* of the Astrolabe, turn it over, and then let the *Rete*
revolve westward till the 1st point of Aries is just within the altitude-
circle marked 25, allowing for the ½ degree by guess. This will bring
the denticle near the letter C, and the first point of Aries near X, which
means 9 A.M. At the same time, the 20th degree of Gemini will be on
the *horizon obliquus*. See fig. 11, Pl. V. This result can be approximately
verified by a common globe thus; elevate the pole nearly 52°; turn the
small brass hour-circle so that the figure XII lies on the equinoctial

which þat I tok for Myn assendent. & in this wyse hadde I the 24
experience for euer-mo in wich maner I sholde knowe the tyde of the
day / & ek myn assendent.   ¶ Tho wold I wyte the same nyght folwyng the howr of the nyght / & wrowhte in this wyse / among an
hep of sterris fixe, it liked me for to take the altitude of the feire 28
white sterre þat is cleped Alhabor / and fond hir sitting on the west
side of the line of Midday, 18 degres of heyhte taken by my rowle
on the bak-side.   ¶ tho sette I the centre of this Alhabor vp-on 18
degrees among myn Almy-kanteras, vp-on the west side ; by cause 32
þat she was fonden on the west side.   tho leide I my label ouer the
degree of the sonne þat was descended vnder the weste Orisonte,
¶ & rikened alle the lettres capitals fro the lyne of Midday vn-to þe
point of my label in the bordure ; ¶ & fond þat it was passed 8 of 36
the clokke the space of 2 degrees / tho loked I doun vp-on myn est
orisonte, ¶ & fond ther 23 degrees of libra assending, whom I tok
for myn assendent ; & thus lerned I to knowe ones for euer in which
Manere I shuld come to the howre of the nyht / and to myn assendent ; 40
as verreyli as may be taken by so smal [an] instrument / but natheles
in general wold I warne the for euere / ne mak the neuere bold to
haue take a Iust Ascendent by thin Astrilabie, or elles to haue sette
Iustly a clokke, whan any celestial body by which þat thow wenest 44
gouer e thilke thynges ben ney the sowth lyne / for trust wel, whan
þat þe sonne is ney the Meridional lyne, the degree of the sonne
rennyth so longe consentrik vp-on the almy-kanteras, þat sothly
thow shalt erre fro the Iust assendent / the same conclusioun sey I by 48
þe centre of any sterre fix by nyht ; and more-ouer, by experience, 1
wot wel that in owre Orisonte, from .xi. of the clokke vn-to on of the

---

colure ; then turn the globe till IX lies under the brass meridian.   In
the next example, by the Astrolabe, let the height of Alhabor (Sirius) be
about 18°.   Turn the denticle Eastward till it touches the 58th degree
near the letter O, and it will be found that Alhabor is about· 18° high
among the *almicanteras*, whilst the first point of Aries points to 32° near
the letter H, i. e. to 8 minutes past 8 P.M ; whilst at the same time,
the 23rd degree of Libra is almost on the *Horizon obliquus* on the Eastern
side.   By the globe, at about 8 min. past 8 P.M., the altitude of Sirius is
very nearly 18°, and the 23rd of Libra is very near the Eastern horizon.
See fig. 12, Plate V.

clokke, in takyng of a Iust assendent in A portatif Astrelabie, hit is
52 to hard to knowe. I mene, from .xi. of the clokke by-forn the howre
of noon til on of the clok next folwyng. ¶ & for the more declaracion, lo her thi figure.

[Fol. 11]   **4. Special declaracion of the assendent.**
          [Specialis declaracio de ascendente.]

¶ The assendent sothly, as wel in alle natiuitez as in questiouns
& elecciouns of tymes, is a thing which þat thise Astrologiens gretly
obseruen / wher-fore me semeth conuenient, sin þat I speke of the
4 assendent, to make of it special declaracioun / The assendent sothly,
to take it at the [largeste], is thilke degree þat assendith at any of
thise forseide tymes v[po]n the est Orisonte; & there-for, yif þat any
planet assende at þat same tyme in thilke for-seide [degre of] his lon-
8 gitude, Men seyn þat thilke planete is in horoscopo. ¶ but sothly, the
hows of the assendent, þat is to seyn, the firste hous or the est Angle,
is a thing more brod & large. ¶ For after the statutz of Astrologiens, what celestial body þat is 5 degres a-boue thilk degre þat
12 assendith, or with-in þat nowmbre, þat is to seyn, nere the degree
þat assendith / yit rikne thei thilke planet in the Assendent. ¶ And
what planete þat is vnder thilke degree þat assendith the space of [25]
degrees, ¶ yit sein thei that thilke planete is lyk to him þat is [in] the
16 hows of the assendent / but sothly, yif he passe the bondes of thise
forseide spaces, a-boue or by-nethe, they sein þat the planete is
failling fro the assendent; / yit sein thise Astrolog[i]ens, that the
assendent ¶ & eke the lord of the assendent, may be shapen for to
20 be fortunat or infortunat / as thus / a fortunat assendent clepen
they whan þat no wykkid planete, as saturne or Mars, or elles the
[* Fol. 11 b.] tail of the dragoun, is in [þe] hows *of the assendent, ne þat
no wikked planete haue non aspecte of enemyte vp-on the assendent;
24 but they wol caste þat thei haue a fortunat planete in hir assendent
& ȝit in his felicite, ¶ & than sey they þat it is wel. ¶ forther-

---

4. The ascendent at any given moment is that degree of the zodiac
which is then seen upon the Eastern horizon. Chaucer says that astrologers reckoned in also 5 degrees *of the zodiac* above, and 25 below,

ouer, they seyn þat the infortunyng of an assendent is the contrarie
of thise forseide thinges.   ¶ the lord of the assendent sey they þat
he is fortunat, whan he is in god place fro the assendent as in angle ; 28
or in a succedent, where-as he is in his dignite & conforted with
frendly aspectys of planetes & [wel] resceiued, & ek that he may sen
the assendent, and þat he be nat retrograd ne combust, ne ioigned
with no shrewe in the same signe / ne that he be nat in his des[c]en- 32
cioun, ne ioigned with no planete in his discencioun, ne haue vp-on him
non aspecte infortunat ; & [than] sey they þat he is wel.   ¶ natheles, theise ben obseruauncez of iudicial matiere & rytes of paiens, in
which my spirit ne hath no feith, ne no knowyng of hir horoscopum; 36
for they seyn þat euery signe is departid in 3 euene parties by [10]
degrees, & thilke porcioun they clepe a face.   ¶ & al-thogh þat a
planete haue a latitude fro the Ecliptik, yit sey [some folk] so þat
the planete arise in þat same signe wyth any degree of the forseide 40
face in which [h]is longitude is rekned, þat yit is the planete in
horoscopo / be it in natiuite or in eleccioun, &c.   ¶ & for the more
declaracioun, lo here the figure.

5. To knowe the verrey equacioun of the degree of the sonne, yif so be þat it falle by-twixe thin Almykanteras.

[Ad cognoscendum veram equacionem de gradu solis, si conti-gerit fore in duas Almicanteras.]

[Fol. 12] ¶ For as moche as the almykanteras in thin astrelabie ben
compownet by two & two, ¶ where-as some Almykanteras in sondri
Astrelabies ben compownet by on and on, or elles by 2 & 2, it is
necessarie to thy lernyng to teche the first to knowe & worke with 4
thin owne Instrument.   ¶ wher-for, whan þat the degree of thy
sonne falleth by-twixe two Almykanteras / or elles yif thin Almykanteras ben grauen with ouer gret a point of a compas / for bothe
thise thinges may causen errour as wel in knowyng of the tid of the 8
day as of the verrey Assendent / thow Most werken in this wise.

---

the object being to extend the planet's influence over a whole " house,"
which is a space of the same length as a *sign*, viz. 30°.   See § 36.

5. This merely amounts to taking the mean between two results.

¶ Set the degree [of] thy sonne vp-on the heyer Almykanter*as* of bothe / & waite wel wher as thin Almury towcheth the bordure, &
12 set ther a prikke of ynke ¶ Set down agayn the degree of thy sonne vp-on the nethere Almy-kanter*as* of bothe / & set ther another prikke. ¶ remewe thanne thin Almury in the bordure euene amiddes bothe prikkes / & this wol lede iustly the degree of thi sonne to
16 sitte by-twixe bothe Almykanter*as* in his riht place. ¶ ley thanne thy label ouer the degree of thy sonne ; & find in the bordure the ver*r*ey tide of the day or of the nyht. ¶ & as verreyly shaltow fynde vp-on thin est orisonte thyn assendent ¶ & for more declara-
20 ciou*n*, lo here thi fig*u*re.

[Fol. 12 *b*.]   **6. To knowe the spring of the dawyng & the ende of the euenyng, the which ben called the two crepusculus :**

[Ad cognoscendum ortum solis & eiūs occasum, que uocatur vulgariter crepusculum.]

¶ Set the nadir of thy sonne vp-on 18 degrees of heyhte / Among thyn Almykanter*as* on the west side, & ley thy label on þe degre of thy sonne, & thanne shal the poynt of thi label schewe the
4 spryng of day. ¶ Also set the nadir of thy sonne vp-on 18 degres of heyhte a-mong thin Almykanter*as* on the [est] side, & ley ouer thy label vp-on the degree of the sonne / & w*it*h the point of thy label fynd in the bordure the ende of the euenyng, þat is, ver*r*ey
8 nyht. the nadir of the sonne is thilke degree þat is opposit to the degree of the sonne, in the 7 signe, as thus / eu*er*y degree of aries bi ordre is nadir to eu*er*y degree of libra by ordre / & taurus to Scorpion / gem*in*i to Sagittare / Cancer to Capriçorne / leo to aquarie /
12 virgo to pisces / & ȝif any degree in thi zodiak be dirk, [h]is nadir*e* shal declare him. ¶ & for the more declaraciou*n*, lo heere thi fig*u*re.

---

6. This depends upon the refraction of light by the atmosphere, owing to which light from the sun reaches us whilst he is still 18° below the horizon.   The nadir of the sun being 18° high on the W. side, the sun itself is 18° below the Eastern horizon, giving the time of dawn ; and if the nadir be 18° high on the E. side, we get the time of the end of the evening twilight. Thus, at the vernal equinox, the sun is 18° high soon after 8 A.M. (roughly speaking), and hence the evening twilight ends soon after 8 P.M., 12 hours later, sunset being at 6 P.M.

7. To knowe the arch of the day, that some folk kallen the day artificial, from the sonne arisyng til hit go to rest[e].

[Ad cognoscendum archum diei, quem vulgus vocat diem artificialem in hoc ab ortu solis vsque ad occasum.]

¶ Set' the degree of thy sonne vp-on thin Est orisonte, & ley thy label on the degree of the sonne, & at the poynt of thy label in the bordure set a prikke. ¶ Turn thanne thi riet aboute til the degree [Fol. 13] of the sonne sit vp-on the west Orisonte, & ley thi label vp- 4 on the same degree of the sonne, & at the point of thi label set a-nother prikke. ¶ rekne thanne the quantite of tyme in the bordure by-twixe bothe prikkes, & tak ther thin ark of the day. ¶ the remenant of the bordure vnder the Orisonte is the ark of the 8 nyht. ¶ thus maistow rekne bothe arches / or euery porcion, of wheither þat the liketh. ¶ & by this Manere of wyrkyng / Maistow se how longe þat any sterre fix dwellith a-boue the erthe, fro tyme þat he risith til he go to reste. ¶ but the day natural, þat is to 12 seyn 24 houris, is the reuolucioun of the equinoxial with as moche partie of the zodiak as the sonne of his propre Moeuinge passeth in the mene while. ¶ & for the more declaracioun, lo her thi figure.

8. To turn the howres in-equales in howres equales.

[Ad conuertendum horas inequales in horas equales.]

¶ Knowe the nombre of the degrees in the howris in-equales, & departe hem by 15, & tak ther thin howris equales. [¶ & for the more declaracioun, lo here thi figure.]

---

7. Ex. The sun being in the 1st point of Cancer on the longest day, its rising will be shewn by the point in fig. 5 where the *horizon obliquus* and *Tropicus Cancri* intersect; this corresponds to a point between P and Q in fig. 2, or to about a quarter to 4 A.M. So too the sunset is at about a quarter past 8, and the length of the day 16½ hours; hence also, the length of the night is about 7½ hours, neglecting twilight.

8. On the same day, the number of degrees in the whole day is about 247¼, that being the number through which the *Rete* is turned in the example to § 7. Divide by 15, and we have 16½ equal hours.

[Fol. 13 b.] **9. To knowe the quantite of the day vulgare, that is to seyen, from spring of the day vn-to verrey nyht.**

[Ad cognoscendum quantitatem diei vulgaris, viz. ab ortu diei vsque ad noctem.]

¶ Know the quantite of thi crepusculis, as I haue tawht in the chapitre by-forn, & adde hem to the arch of thi day artificial / & tak ther the space of alle the hole day vulgar, vn-to verrey nyht. / The 4 same manere maistow worke to knowe the quantite of the vulgar nyht. / & for the more declaracioun, lo here the figure.

10. To knowe the quantite of howres in-equales by day.

[Ad cognoscendum horas inequale[s] in die.]

¶ Vnderstond wel, þat thise howris in-equalis ben cleped howres of planetes, & vnderstond wel þat som tyme ben thei lengere by day [than] by nyht, & som tyme the contrarie. ¶ but vnderstond wel 4 þat euermo generaly þe howr in-equal of the [day with þe howr in-equal of the] nyght contenen [30] degrees of the bordure, whiche bordure is euer-mo answering to the degrees of the equinoxial; wher-for departe the arch of the day artificial in 12, & tak ther the quan-8 tite of the howr in-equal by day. ¶ & ʒif thow abate the quantite of the howr in-equal by daye owt of 30 / than shal the remenant þat leueth performe the howr inequal by nyght. ¶ & for the more declaracioun, lo here the figure.

[Fol. 14] **11. To knowe the quantite of howres equales.**

[Ad cognoscendum quantitatem horarum inequalium.]

¶ The quantite of howres equales, þat is to seyn, the howres of the clokke / ben departid by 15 degrees al-redy in the bordure

---

9. The "day vulgar" is the length of the "artificial day," with the length of the twilight, both at morn and at eve, added to it.

10. If, as in § 7, the day be 16½ hours long, the length of each "hour inequal" is 1 h. 22½ m.; and the length of each "hour inequal" of the night is the 12th part of 7½ hours, or 37½ m.; and 1 h. 22½ m., added to 37½ m., will of course make up 2 hours, or 30°.

of thin astralabie, as wel by nyht as by day, generaly for euere.
¶ What nedith more declaracioun? ¶ Wher-for, whan the list to 4
know how manye howres of the clokke ben passed, or any part of any
of thise howris þat ben passed, ¶ or elles how many howres or partie
of howres ben to come, fro swich a tyme to swych a tyme, by day or
by nyhte, ¶ knowe the degree of thy sonne, & ley thy label on it / 8
turne thi Riet abowte ioyntly with thy label, & with the point of it
rekne in the bordure fro the sonne arise vn-to the same place ther
thow desirest, by day as by nyhte / this conclusioun wol I declare in
the laste chapitre of the [4] partie of this tretis so openly, þat [þer] 12
shal lakke no worde þat nedith to the declaracioun. ¶ & for the
more declaracioun, lo here the figure.

12. Special declaracioun of the howres of planetes.
Specialis declaracio de horis planetarum.]

¶ Vnderstond wel, þat euere-mo fro the arising of [the] sonne til
it go to reste / the nader of the sonne shal shewe the howr of the
planete / & fro that tyme forward / al the nyht til the sonne arise /
than shal the verrey degree of the sonne shewe the howr of the 4
planete. Ensample as thus. ¶ the xiij. day of March fil vp-on a
saterday per auenture / & at the arising of the sonne, I fond the
secounde degree of aries sitting vp-on myn est Orisonte, al be it þat
it was but lite ; *than fond I the [2] degree of libra, nadir of [* Fol. 14 b.] 8
my sonne, dessending on my west Orisonte, vp-on which west Orisonte euery day generally, at the sonne ariste, entrith the howr of any
planete, after which planete the day berith his name ; ¶ & endith in

---

11. This merely repeats that 15° of the border answer to an hour of
the clock.
12. This "hour of the planet" is a mere astrological supposition, involving no point of astronomy. Each hour is an "hour inequal," or the
12th part of the artificial day or night. The assumptions are so made
that the *first* hour of every day may resemble the *name of the day ;* the
first hour of Sunday is the hour of the *Sun,* and so on. These hours may
be easily found by the following method. Let 1 represent both Sunday
and the Sun ; 2, Monday and the Moon ; 3, Tuesday and Mars; 4, Wednes-
day and Mercury ; 5, Thursday and Jupiter ; 6, Friday and Venus ; 7,

12 the nexte strik of the plate vnder the forscide west Orisonte / & euere
as the sonne clymbith vppere & vppere, so goth his nadir downere
& downere, techyng by swych strikes the howres of planetes by ordre
as thei sitten in the heuene.   the first howr inequal of euery Sat-
16 terd[a]y is to Saturne; ¶ & the secounde to Iupiter; ¶ the 3 to
Mars; the 4 to the sonne; ¶ the 5 to venus; ¶ the 6 to Mercurius;
¶ the 7 to the mone; ¶ & thanne agayn the 8 is to saturne; ¶ the
9 to Iupiter; ¶ the 10 to Mars; ¶ the 11 to the sonne; ¶ the 12
20 to venus; ¶ And now is my sonne gon to reste as for that setter-
day.   ¶ Thanne shewyth the verrey degree of the sonne the howr
of Mercurie entryng vnder my west orisonte at eue; ¶ & next him
succedith the Mone; ¶ & so forth by ordre, plancte aftur planete,
24 in howr after howr, al the nyht longe til the sonne arise.   ¶ now
risith the sonne þat Sonday be the morwe; ¶ & the nadir of the sonne
vp-on the west Orizonte shewith me the entring of the howre of the
forseide sonne.   ¶ & in this maner succedith planete vnder planete,
28 fro saturne vn-to the mone, & fro the mone vp a-gayn to satourne,
howre after howre generaly.   ¶ & thus know[e] I this conclusioun.
¶ & for the more declaracioun, lo here the figure.

Saturday and Saturn.   Next, write down the following succession of
figures, which will shew the hours at once.

1642753|164275316427535316427531631642753164275316.

Ex. To find the planet of the 10th hour of Tuesday.   Tuesday is the
third day of the week; begin with 3, to the left of the upright line, and
reckon 10 onwards; the 10th figure (counting 3 as the *first*) is 6, i. e.
Venus. So also, the planet of the 24th hour of Friday is the Moon, and
Saturday begins with Saturn. It may be observed that this table can be
carried in the memory, by simply observing that the numbers are written,
beginning with 1, in the *reverse order of the spheres,* i. e. Sun, Venus,
Mercury, Moon; and then (beginning again at the outmost sphere) Saturn,
Jupiter, Mars. This is why Chaucer takes a *Saturday;* that he may begin with the remotest planet, *Saturn,* and follow the reverse order of the
spheres. See fig. 10, Pl. V. Here, too, we have the obvious reason for
the succession of the names of the days of the week, viz. that the plancts
being reckoned in this order, we find the Moon in the 25th place or hour
from the Sun, and so on.

[Fol. 15] **13. To knowe the altitude of the sonne in Middes of the day, that is cleped the altitude Meridian.**

[Ad cognoscendum altitudinem solis in medio diei, que vocatur altitudo meridiana.]

¶ Set the degree of the sonne vp-on the lyne Meridional, & rikene how many degrees of Almykanteras ben by-twyxe thyn est Orisonte & the degree of the sonne. ¶ & tak ther thyn altitude Meridian / this [is] to seyne, the heiest of the sonne as for that day. ¶ So maistow knowe in the same lyne, the heiest cours þat any sterre fix clymbith by nyht; ¶ this is to seyn, þat w[h]an any sterre fix is passed the lyne Meridional, than by-gynnyth it to descende, & so doth the sonne. ¶ & for the more declaracioun, lo here thi figure.

14. To knowe the degree of the sonne by thy riet, for a maner curiosite, &c.

[Ad cognoscendum gradum solis curiose.]

¶ Sek bysily with thi rewle the heiest of the sonne in Midde of the day ; ¶ turne thanne thyn Astrelabie, & with a prikke of ynk marke the nombre of þat same Altitude in the lyne Meridional. turne thanne thy Ryet a-bowte til thow fynde a degree of thi zodiak acording with the prikke, ¶ this is to seyn, sittynge on the prikke ; ¶ & in soth, thow shalt fynde but 2 degrees in al the zodiak of that condicioun ; ¶ & yit thilke 2 degrees ben in diuerse signes ; ¶ than maistow lyhtly by the sesoun of the yere knowe the signe in whiche þat is the sonne. [¶ & for the more declaracioun, lo here thi figure.]

---

13. The reason of this is obvious from what has gone before. The sun's meridional altitude is at once seen by placing the sun's degree on the South line.

14. This is the exact converse of the preceding. It furnishes a method of testing the accuracy of the drawing of the almikanteras.

[Fol. 15 b.] **15. To know which day is lik to wych day as of lengthe, &c.**

[Ad cognoscendum quales dies in longitudine sunt similes.]

¶ Loke whiche degrees ben illik fer fro the heuedes of Cancer & Capricorn͞; & lok, whan the sonne is in any of thilke degrees, than ben the dayes ilike of lengthe. ¶ this is to seyn, þat as long 4 is þat day in þat Monthe, as was swych a day in swich a month ¶ ther variet[h] but lite. ¶ Also yif þow take 2 daies naturaly in the yer ilike fer fro eyther pointe of [the] equinoxial in the opposit parties, ¶ than as long is the day artificial of þat on day / as is the 8 nyht of þat othere, & the contrarie. ¶ & for [the] more declaracioun, lo here thi figure.

16. This chapitre is a Maner declaracioun to conclusiouns þat folwen.   ⁚

[Illud capitulum est quedam declaracio ad certas conclusiones sequentes.]

¶ Vnderstond wel þat thy zodiak is departid in 2 halfe cercles, as fro the heued of capricorne vn-to the heued of Cancer / & agaynward fro the heued of cancer vn-to the heued of Capricorne. 4 ¶ the heued of Capricorne is the lowest point, wher as the sonne goth in wynter; ¶ & the heued of Cancer is the heiest point, in whiche the sonne goth in somer. ¶ & ther-for vnderstond wel, þat any two degrees þat ben ilike fer fro any of thise two heuedes / truste 8 wel þat thilke two degrees ben of il[i]ke declinacioun, be it sowth-ward or northward ; & the daies of hem ben ilike of lengthe, & the

---

15. This is best done by help of the *back* of the instrument, fig. 1. Thus May 13 (old style), which lies 30° to the W. of the S. line, is nearly of the same length as July 13, which lies 30° to the E. Secondly, the day of April 2 (old style), 20° above the W. line, is nearly of the same length as the night of Oct. 2, 20° below the E. line, in the opposite point of the circle. This is but an approximation, as the divisions on the in-strument are rather minute.

16. This merely expresses the same thing, with the addition, that on days of the same length, the sun has the same meridional altitude, and the same declination from the equator

nyhtes also ; & the shadwes ilike, & the Altitudes ilike at Midday for eue*re*.   ¶ & for more declaraciou*n*, lo here thi figu*re*.

[Fol. 16] **17. To knowe the verrey degree of any maner sterre straunge or vnstraunge after his longitude, thow he be inde-terminat in thin astralabie ; sothly to the trowthe, thus he shal be knowe.**

[Ad cognoscendum verum gradum alicuius stelle aliene secundum eius latitudinem (*sic*), quamvis sit indeterminata in astro-labio ; veraciter isto modo.]

¶ Tak the altitude of this sterre whan he is on the Est side of the lyne Me*ri*dional, as ney as thow maist gesse ; ¶ & tak an assendent a-non riht by som mane*r* sterre fix which þat þow knowest ¶ & for-get nat the altitude of the firste sterre, ne thyn assendent ; 4
& whan þat this is don / espie diligently whan this same firste sterre passeth any-thing the sowth westward, and hath him a-non riht in the same nowmbre of altitude on the west side of this lyne Me*ri*dional ¶ as he was kawht on the est side ; & tak a newe assendent a-non 8
riht by som Mane*r* sterre fixe which þat thow knowest ; & for-get nat this secounde assendent. ¶ and whan þat this is don, rikne thanne how manye degrees ben by-twixe the firste assendent & the seconde assendent / & rikne wel the Middel degree by-twyne bothe Assend-12

---

17. Here *passeth any-thing the sowth westward* means, passes some-what to the westward of the South line.   The problem is, to find the de-gree of the zodiac which is on the meridian with the star.   To do this, find the altitude of the star *before* it souths, and by help of problem 3, find out the ascending degree of the zodiac ; secondly, find the ascending degree at an equal time *after* it souths, when the star has the same alti-tude as before, and the mean between these will be the degree that ascends when the star is on the meridian.   Set this degree upon the Eastern part of the *horizon obliquus*, and then the degree which is upon the meridional line souths together with the star.   Such is the solution given, but it is but a very rough approximation, and by no means always near to the truth.   An example will shew why.   Let Arcturus have the same altitude at 10 P.M. as at 2 A.M.   In the first case the 4th of Sagit-tarius is ascending, in the second (with sufficient accuracy for our pur-pose) the 2nd of Aquarius ; and the mean between these is the 3rd of Capricorn.   Set this on the Eastern horizon upon a globe, and it will be

entes, & set thilke Middel degree vp-on thin est Orisonte ; ¶ &
waite thanne what degre þat sit vp-on the line Meridional, & tak ther
the verrey degre of the Ecliptik in which the sterre stondeth for
16 the tyme.   ¶ For in the Ecliptik is the longitude of a celestial body
rekened, euene fro the heued of aries vn-to [the] ende of pisces.
¶ & his latitude is Rikned after the quantite of [h]is declinacion,
north or sowth to-warde þe poles of this world / as thus. ¶ yif it be
20 of the sonne or of any fix sterre / rekene [h]is latitude or his declinacioun fro the Equinoxial cercle ; ¶ & yif it be of a planete, rekne
than the quantite of [h]is latitude fro the Ecliptik lyne.   ¶ Al be it
so þat fro [the] Equinoxial may the declinacion or the latitude of any
24 body celestial be rikned, after the site north or south, & after the
quantite of [h]is declinacion. ¶ & riht so may the latitude or the declinacion of any body celestial, saue only of the sonne, after his site
north or south, & after the quantite of his declinacioun, be rekned fro
28 the Ecliptik lyne ;   ¶ Fro which lyne alle planetes som tyme declinen
north or south, saue only the for-seide sonne. ¶ & for the more
declaracioun, lo here thi figure.

---

seen that it is 20 min. past midnight, that 10° of Scorpio is on the meridian, and that Arcturus has past the meridian by 5°. At true midnight,
the ascendent is the 29° of Sagittarius. The reason of the error is that
right ascension and longitude are here not sufficiently distinguished.
By observing the degrees of the *equinoctial*, instead of the *ecliptic*, upon
the Eastern horizon, we have at the first observation 272°, at the second
332°, and the mean of these is 302°; from this subtract 90°, and the result, 212°, gives the right ascension of Arcturus very nearly, corresponding to which is the beginning of the 5° of Scorpio, which souths along
with it. This latter method is correct, because it assumes the motion
to take place round the axis of the equator. The error of Chaucer's method is that it identifies the motion of the equator with that of the
ecliptic. The amount of the error varies considerably, and may be rather
large. But it can easily be diminished, (and no doubt was so in practice),
by taking the observations *as near the south line as possible.* Curiously
enough, the rest of the section explains the difference between the two
methods of reckoning. The modern method is to call the co-ordinates
right ascension and *declination*, if reckoned from the equator, and *longitude* and *latitude*, if from the ecliptic. Motion in *longitude* is not the
same thing as motion in *right ascension.*

[Fol. 16 b.] 18. To knowe the degrees of the longitudes of fixe sterres after þat they ben determinat in thin astralabie, yif so be þat they ben trewly set.

[Ad cognoscendum gradus longitudinis de stellis fixis que determinantur in astrolabio, sicut in suis locis recte locentur.]

¶ Set the centre of the sterre vp-on the lyne Meridional, & tak kep of thi zodiak, ¶ & loke what degree of any signe þat sit on the same lyne Meridional at þat same tyme, & tak the degree in which the sterre standith ; ¶ & with that same degree com[e]th þat same 4 sterre vn-to that same lyne fro the Orisonte. ¶ & for more declaracioun, lo here thi figure.

19. To knowe with which degree of the zodiak any sterre fixe in thin Astrelabie arisith vp-on the est Orisonte, Al-they [h]is dwellyng be in a-nother signe.

[Ad cognoscendum cum quibus gradibus zodiaci que stella fyxa in astrolabio ascendit super orizontem orientalem, quamuis eius orizon (sic) sit in alio signo.]

¶ Set the Sentre of the sterre vp-on the est Orisonte, ¶ & loke what degre of any signe þat sit vp-on the same Orisonte at þat same tyme. ¶ And vnderstond wel, þat with þat same degre arisith þat same sterre ; and thys *merueyllous arising with a strange de- [* Fol. 17] 4 gree in another signe is by-cause þat the latitude of the sterre fix is either north or sowth fro the equinoxial. ¶ but sothly, the latitudes

---

18. The "centre" of the star is the technical name for the extremity of the metal tongue representing it. The "degree in which the star standeth" is considered to be that degree of the zodiac which souths along with it. Thus Sirius or Alhabor has its true longitude nearly equal to that of 12° of Cancer, but, as it souths with the 9th degree, it would be said to stand in that degree. This may serve for an example ; but it must be remembered that its longitude was different in the time of Chaucer.

19. Also it rises with the 19th degree of Leo, as it is at some distance from the zodiac in latitude. The same "marvellous arising in a strange sign" is hardly because of the latitude being north or south from the *equinoctial*, but rather because it is north or south of the *ecliptic*. For example, Regulus (a Leonis) is on the ecliptic, and of courses rises

of planetes ben comu[n]ly rekned fro the Ecliptik, bi-cause þat non
8 of hem declinet[h] but fewe degrees owt fro the brede of the zodiak.
¶ & tak god kep of this chapitre of arising of the celestial bodies;
for truste wel, þat neyther mone ne sterre as in owre Embelif
orisonte arisith *with* þat same degree of his longitude, saue in O cas;
12 ¶ & that is, whan they haue no latitude fro the Ecliptik lyne. but
natheles som tyme is cueriche of thes planetes vnder the same lyne.
¶ & for more declaracioun, lo here thi figure.

20. To knowe the declinacioun of any degree in the zodiak fro the equinoxial cercle, &c.

[**Ad cognoscendum declinacionem alicuius gradus [in] zodiaco ⌊a⌋ circulo equinoctiali.**]

¶ Set the degree of any signe vp-on the lyne Meridional, &
rikne [h]is altitude in Almykanteras fro the Est Orizonte vp to the
same degree set in the forseide lyne, & set ther a prikke. ¶ turne vp
4 thanne thy Riet, and set the heued of aries or libra in the same
Meridional lyne, & set ther a-nother prikke. ¶ & whan þat this is
[* Fol. 17 b.] don, considere the *Altitudes of hem bothe; for sothly the
difference of thilke altitudes is the declinacion of thilke degre fro
8 the equinoxial. ¶ & yif so be þat thilke degree be northward fro the
equinoxial, than is his declinacion north; ¶ yif it be sowthward,
than is it sowth. ¶ & for the more declaracioun, lo here thi figure.

---

with that very degree in which it is. Hence the reading *equinoctial*
leaves the case in doubt, and we find a more correct statement just below, where we have "whan they haue no latitude fro the Ecliptik lyne."
At all places, however, upon the earth's equator, the stars will rise with
the degrees of the zodiac in which they stand.

20. Here the disc (fig. 5) is supposed to be placed beneath the Rete
(fig. 2). The proposition merely tells us that the difference between the
meridian altitudes of the given degree of the zodiac and of the 1st point
of Aries is the *declination* of that degree, which follows from the very definition of the term. There is hardly any necessity for setting the second
prick, as it is sufficiently marked by being the point where the equinoctial circle crosses the south line. If the given degree lie *outside* this
circle, the declination is *south*; if *inside*, it is *north*.

21. To knowe for what latitude in any regioun the **Almikanteras** of any table ben compowned.

[Ad cognoscendum pro qua latitudine in aliqua regione almicantre tabule mee sunt composite.]

¶ Rikne how manie degrees of Almikanter*as* in the M*e*ridional lyne be fro the cercle equinoxial vn-to the senyth ; ¶ Or elles fro the pool artik vn-to the north Orisonte ; & for so gret a latitude or for so smal a latitude is the table compowned. ¶ & for more declaracion, lo here thi fig*ur*e.

[Fol. 18] 22. To knowe in special the latitude of owre countray, I mene after the latitude of **Oxenford,** & the heyhte of owre pol.

[Ad cognoscendum specialiter latitudinem nostri centri (*sic*), scilicet latitudinem Oxonie, et altitudinem poli nostri.]

Vnderstond wel, þat as fer is the heued of aries or libra in the equinoxial from owre orisonte as is the cenyth fro the pole artik ; ¶ & as hey [is] the pol Artik fro the Orisonte as the Equinoxial is fer fro the senyth. ¶ I proue it thus by the latitude of Oxenford ; vnderstond wel, þat the heyhte of owre pool Artik fro owre north Orisonte is 51 degrees & 50 Minutes ; than is the cenyth from owre pool Artik 38 degrees & 10 Minutes ; than is the equinoxial from owre senyth 51 degrees & 50 Minut*es*; ¶ than is owre south Orisonte from owre equinoxial 38 degrees & 10 Minutes. ¶ vnderstond wel

---

21. In fig. 5, the almicanteras, if accurately drawn, ought to shew as many degrees between the south point of the equinoctial circle and the zenith as are equal to the latitude of the place for which they are described. The number of degrees from the pole to the northern point of the *horizon obliquus* is of course the same. The latitude of the place for which the disc is constructed is thus determined by inspection.

22. In the *first* place where "*orisonte*" occurs, it means the *South* point of the horizon ; in the *second* place, the *North* point. By referring to fig. 13, Plate V, it is clear that the arc ♈S, representing the distance between the equinoctial and the S. point is equal to the arc ZP, which measures the distance from the pole to the zenith ; since PO♈ and ZOS are both right angles. Hence also Chaucer's second statement, that the arcs PN and ♈Z are equal. In his numerical example, PN is 51° 50' ;

this Reknyng. ¶ Also for-get nat þat the cenyth is 90 degrees of heyhte fro the Orisonte, & owre equinoxial is 90 degrees from owre 12 pool Artik. ¶ Also this shorte rewle is soth, þat the latitude of any place in a regioun is the distance fro the senyth vnto the Equinoxial. ¶ & for more declaracioun, lo here þi figure.

23. To proue euidently the latitude of any place in a Regioun, by the preue of-the heyhte of the pol Artik in þat same place.

[Ad probandum euidenter latitudinem alicuius loci in aliqua regione, per probacionem altitudinis de polo artico in eodem loco.]

In some wynters nyht, whan the firmament is clere & thikke-sterred / waite a tyme til þat any sterre fix sit lyne-riht perpendiculer [* Fol. 18 b.] ouer * the pol Artik, & clepe þat sterre A. ¶ & wayte a-nother 4 sterre þat sit lyne-riht vnder A, & vnder the pol / & clepe þat sterre F. And vnderstond wel, þat F is nat consideret but only [to] declare þat A sit euene ouere the pool. ¶ tak thanne a-non riht the altitude of A from the Orisonte / & forget it nat. ¶ Lat A & F go far-8 wel til agayns the dawenyng a gret while / & come thanne agayn & Abid til þat A is euene vnder the pol & vnder F ; ¶ for sothly, than wole F sit ouer the pool / & A wol sit vnder the pool. ¶ tak than eft-sones the altitude of A from the Orisonte ¶ & note as wel [h]is 12 secounde altitude as his furste Altitude / & whan þat this is don, ¶ rikne how manye degrees þat the firste altitude of A excedith his seconde altitude, ¶ & tak half thilke porcioun þat is excedit, & adde it to his seconde altitude ; ¶ & tak ther the eleuacioun of thi pool, 16 & eke the latitude of thy regioun ; ¶ for thise two ben of a nombre ; ¶ .this is to seyn, as many degrees as thy pool is eleuat / so michel

---

and therefore ZP is the complement, or 38° 10′. So also ♈Z is 51° 50′ ; and ♈S is 38° 10′. Briefly, ♈Z measures the latitude.

23. Here the altitude of a star (A) is to be taken twice ; firstly, when it is on the meridian in the most *southern* point of its course, and secondly, when on the meridian in the most *northern* point, which would be the case twelve hours later. The mean of these altitudes is the altitude of the pole, or the latitude of the place. In the example given, the star A is only 4° from the pole, which shews that it is the Pole-star, then farther

is the latitude of the Regiou*n*.   ¶ Ensample as thus : p*ar* auenture
the altitude of A in the euenyng is 56 degrees of heyhte ¶ than wol
his seconde altitude or the dawyng be 48 / þat is [8] lasse than 56, 20
þat was his furste altitude at euen. ¶ tak*e* thanne the half of 8 / &
adde it to 48, þat was [h]is second*e* altitude, and [than] hastow 52.
¶ now hastow the heyhte of thy pol and the latitude of the regiou*n*.
¶ but vnderstond wel þat to proue this conclusiou*n* & many a nother 24
fair conclusiou*n*, thow most haue a plomet hanging on a lyne heyer
than thin heued on a p*er*che ¶ & thilke lyne mot hange euene p*er*pen-
diculer by-twixe the pool & thin eye / & thanne shal-tow sen yif A
sitte euene ouer the pool & ou*er* F at euene / & also yif F sitte euene 28
ou*er* the pool & ou*er* A or day. ¶ & for more declaraci*o*n, lo here
thi fig*u*re.

[Fol. 19] **24. Another conclusioun to proue the heyhte of the pool
Artik fro the orisonte.**

**[Alia conclusio ad probandum altitudinem de polo artico ab
orizonte.]**

Tak any sterre fixe þat neu*er*e dissendith vnder the Orisonte in
thilke regiou*n*, & considere his heiest altitude & his lowest Altitude
fro the Orisonte ; ¶ & make a nombre of bothe thise altitudes ; tak
thanne & abate half þat nombre, & tak þer the eleuaciou*n* of the pol 4
Artik in þat same Regiou*n* / & for more declaraciou*n*, lo here thi
fig*u*re.

from the Pole than it is now.   The star F is, according to Chaucer, any
convenient star having a right ascension differing from that of the Pole-
star by 180° ; though one having the *same* right ascension would serve
as well.   If then, at the first observation, the altitude of A be 56, and
at the second be 48, the altitude of the pole must be 52.   See fig. 13,
Plate V.

24. This comes to much the same thing.   The *lowest* or northern
altitude of Dubhe (a Ursæ Majoris) may be supposed to be observed to be
25°, and his *highest* or southern altitude to be 79°.   Add these ; the sum
is 104 ; "abate" or subtract half of that number, and the result is 52° ;
the latitude.

25. A-nother conclusioun to proue the latitude of the Regioun, &c.
[Alia conclusio ad probandum latitudinem regionis.]

¶ Vnderstond wel þat the latitude of any place in A Regioun is
verreyly the space by-twixe the senythe of hem þat dwellen there &
[the] equinoxial cerkle, north or sowthe, takyng the mesure in the
4 Meridional lyne, as shewith in the Almykanteras of thin Astrelabie.
¶ & thilke space is as moche as the pool artik is hey in the same
place fro the Orisonte. ¶ And than is the depressioun of the pol
antartik, þat is to seyn, than is the pol antartik by-nethe the
8 Orisonte the same quantite of space, neither mor ne lasse. thanne,
yif thow desire to knowe this latitude of the Regioun, tak the altitude of the sonne in the Middel of the day, whan the sonne is in the
heuedes of aries or of libra / for thanne Moeuyth the sonne in the
12 lyne equinoxial ; ¶ & abate the nombre of that same sonnes Altitude
[* Fol. 19 b.] owt of 90, & thanne is the remenaunt * of the noumbre þat
leuyth the latitude of the Regioun, as thus : I suppose that the sonne
is thilke day at noon 38 degrees And 10 minutes of heyhte. Abate
16 thanne thees degrees And minutes owt of 90 ; so leueth there 51
degrees and 50 minutes, the latitude. ¶ I sey nat this but for ensample ; for wel I wot the latitude of Oxenforde is certein minutes
lasse, as y myght proue. ¶ Now yif so be þat the semith to long a
20 tarienge, to abide til þat [þe] sonne be in the heuedes of aries or of
libra, thanne whaite whan the sonne is in any other degree of the
zodiak, & considere the degree of [h]is declinacion fro the equinoxial
lyne ; ¶ & yif it so be þat the sonnes declinacion be northward fro the
24 equinoxial, abate thanne fro the sonnes altitude at noon the nombre

---

25. Here, as in § 22, Chaucer says that the latitude can be measured
by the arc Z♈ or PN ; he adds that the depression of the Antarctic pole,
viz. the arc SP′ (where P′ is the S. pole), is another measure of the
latitude. He explains that an obvious way of finding the latitude is by
finding the altitude of the sun at noon at the time of an equinox. If
this altitude be 38° 10′, then the latitude is the complement, or 51° 50′.
But this observation can only be made on two days in the year. If
then this seems to be too long a tarrying, observe his midday altitude,
and allow for his declination. Thus, if the sun's altitude be 58° 10′ at
noon when he is in the first degree of Leo, subtract his declination, viz.
20°, and the result is 38° 10′, the complement of the latitude. If, how-

of his declinac*i*on, ¶ & thanne hastow the heyhte of the heuedes of aries & lib*r*a; as thus / My sonne is, p*ar* Auent*u*re, in þe firste degre of leou*n*, 58 degrees and 10 Minutes of heyhte at noon / & his declinac*i*on is almost 20 degrees northward fro the equinoxial; 28 abate thanne thilke 20 degrees of declinac*i*on owt of the altitude at noon, than leueth the 38 degrees and odde Minutes; lo ther the heued of aries or libra, & thin equinoxial in that Regiou*n*.   ¶ Also yif so be þat the sonnes declinaciou*n* be sowthward fro the Equi- 32 noxial, ¶ Adde thanne thilke declinac*i*on to the altitude of the sonne at noon / and tak ther the heuedes of aries & libra & thin Equinoxial. ¶ abate thanne the heyhte of the Equinoxial owt of 90 degrees, & thanne leuyth there þe distans of the pole, 51 degrees & 50 36 Minutes, of that regiou*n* fro the Eq*u*inoxial.   ¶ Or elles, yif the lest, take the heiest altitude fro the equinoxial of any sterre fix that thow knowest, & tak his nethere elongaciou*n* lengthing fro the same eq*u*inoxial lyne, & wirk*e* in the man*er* forseid. ¶ & for more declara- 40 c*i*on, lo here thi fig*ur*e.

[Fol. 20] **26. Declaracioun of the assensioun of signes, &c.**

[Declaracio de ascensione signorum.]

The excellence of the sp*er*e solide, amonges other noble conclusiou*n*s, shewyth Manifeste the diu*er*se assenciou*n*s of signes in diu*er*se places, as wel in the rihte cercle as in the Embelif cercle. ¶ thise Auctours writen þat thilke signe is cleped of riht Ascensiou*n*, 4 wi*th* which more p*ar*t of the cercle Eq*u*inoxial & lasse p*ar*t of the zodiak ascendith / & thilke signe assendith Embelif, wi*th* whiche lasse p*ar*t of the Eq*u*inoxial and more p*ar*t of the zodiak assendith.

---

ever, the sun's declination be *south*, the amount of it must be added instead of subtracted.   Or else we may find ♈F′, the highest altitude of a star F′ above the equinoctial, and also ♈F, its nether elongation extending from the same, and take the mean of the two.

26. The "Sphere Solid" answers nearly to what we now call a globe. By help of a globe it is easy to find the ascensions of signs for *any latitude*, whereas by the astrolabe we can only tell them for those latitudes for which the plates bearing the almicanteras are constructed.   The signs which Chaucer calls "of right (i. e. direct) ascension" are those signs of

8 [Ferther ouer they seyn, that in thilke cuntrey where as the senith
of hem that dwellen there is in the equinoxial lyne and her orisonte
passyng by the poles of this worlde, thilke folke han this right cercle
and the right orisonte ;] ¶ & euere mo the Arch of the day & the arch
12 of the nyht is ther ylike long, & the sonne twyes euery yer passinge
thorow the cenyth of her heued ; & 2 someres & 2 wynteres in a yer
han this forseide poeple. ¶ And the Almykanteras in her Astrolabies
ben streyhte as a lyne / so as shewyth in this figure. ¶ The vtilite to
16 knowe the Assenciouns in the rihte cercle is this / truste wel þat by
mediacioun of thilke assenciouns thise Astrologiens by hir tables &

---

the zodiac which rise more directly, i. e. at a greater angle to the horizon
than the rest. In latitude 52°, Libra rises so directly that the whole
sign takes more than $2\frac{3}{4}$ hours before it is wholly above the horizon,
during which time nearly 43° of the equinoctial circle have arisen ; or,
in Chaucer's words, "the more part." (i. e. a larger portion) of the equinoctial ascends with it. On the other hand, the sign of Aries ascends
so obliquely that the whole of it appears above the horizon in less than
an hour, so that a "less part" (a smaller portion) of the equinoctial
ascends with it. The following is a rough table of Direct and Oblique
Signs, shewing approximately how long each sign takes to ascend, and
how many degrees of the equinoctial ascend with it, in lat. 52°.

Oblique Signs.	Degrees of the Equinoctial.	Time of ascending.	Direct Signs.	Degrees of the Equinoctial.	Time of ascending.
Capricornus	26°	1 h. 44 m.	Cancer	39°	2 h. 36 m.
Aquarius	16°	1 h. 4 m.	Leo	42°	2 h. 48 m.
Pisces	14°	0 h. 56 m.	Virgo	43°	2 h. 52 m.
Aries	14°	0 h. 56 m.	Libra	43°	2 h. 52 m.
Taurus	16°	1 h. 4 m.	Scorpio	42°	2 h. 48 m.
Gemini	26°	1 h. 44 m.	Sagittarius	39°	2 h. 36 m.

These numbers are sufficiently accurate for the present purpose.

In l. 8, there is a gap in the sense in nearly all the MSS., but the
Bodley MS. 619 fortunately supplies what is wanting, to the effect
that, at places situated on the equator, the poles are in the horizon. At
such places, the days and nights are always equal. Chaucer's next
statement is true for *all* places *within the tropics*, the peculiarity of
them being that they have the sun vertical twice in a year. The
statement about the "two summers and winters" is best explained
by the following. "In the tropical climates, . . seasons are caused
more by the effect of the winds (which are very regular, and depend
mainly on the sun's position) than by changes in the direct action of
the sun's light and heat. The seasons are not a summer and winter,
so much as recurrences of wet and dry periods, *two in each year*."
—English Cyclopædia; *Seasons, Change of.* Lastly, Chaucer reverts to
places on the equator, where the stars all seem to move in vertical

ƀir instrumentz knowen verreyly the Assencioun of euery degree &
Mynut in al the zodiak, as shal be shewyd.   ¶ And nota, þat this
forseid rihte orisonte, þat is clepid *orison rectum* / diuideth the 20
equinoxial in-to riht Angles ; & the embelif orisonte, wher as the
pol is enhawsed vp-on the orisonte, ouerkeruyth the equinoxial in
Embelif Angles, as shewith in the figure.   ¶ & for the more declara-
cioun, lo here the figure.   24

27. This is the conclusioun to knowe the Assenciouns of signes in the riht cercle, þat is, circulus directus, &c.

[Ad cognoscendum ascenciones signorum in recto circulo, qui
vocatur circulus directus.]

[Fol. 20 b.] Set the heued of what signe the liste to knowe his
Assending in the riht cercle / vp-on [the] lyne Meridional, ¶ & waite
wher thin Almury towchiet[h] the bordure, & set ther a prikke / turne
thanne thy riet westward til þat the ende of the forseide signe sitte 4
vp-on the Meridional lyne / & eft-sones waite wher thin almury
towchith the bordure, & set¹ ther Another prikke.   Rikne thanne the
nombre of degrees in the bordure by-twyxe bothe prikkes, ¶ & tak
the Assencioun of the signe in the riht cercle.   And thus maistow 8
wyrke with euery porcioun of thy zodiak, &c.   ¶ & for the more
declaracioun, lo her thi figure.

28. To knowe the assencions of signes in the Embelif cercle in euery regioun, I Mene, in circulo obliquo.

[Ad cognoscendum ascenciones signorum in recto (*sic !*) circulo
in omni regione, hoc est, in circulo obliquo.]

¶·Set the heued of the signe which as the list to knowe his As-

---

circles, and the almicanteras are therefore straight lines.   The line
marked *Horizon Rectus* is shewn in fig. 5, where the *Horizon Obliquus*
is also shewn, cutting the equinoctial circle obliquely.
   27. The real object in this section is to find how many degrees of the
equinoctial circle pass the meridian together with a given zodiacal sign.
Without even turning the *rete*, it is clear that the sign Aries, for in-
stance, extends through 28° of the equinoctial ; for a line drawn from
the centre, in fig. 2, through the end of Aries will (if the figure be cor-
rect) pass through the end of the 28th degree below the word *Oriens*.
   28. To do this accurately requires a very carefully marked Astro-

censioun vp-on the est Orisonte, ¶ & waite wher thyn Almury towchith the bordure, & [set] ther a prikke.   ¶ turne thanne thy riet
4 vpward til þat the ende of the same signe sitte vp-on the Est Orisonte / and waite eft-sones wher as thin almury towcheth the
bordure, & set ther a-nother prikke.   ¶ Rikne thanne the nowmbre
of degrees in the bordure by-twyxe bothe prikkes, & tak ther the
8 Assencioun of the signe in the Embelif cercle.   ¶ & vnderstond wel,
þat alle signes in thy zodiak fro the heued of aries vnto the ende
of virgo ben cleped signes of the north fro the Equinox[i]al, ¶ &
the[se] signes arisen by-twyxe the verrey est & the verrey north in
12 owre Orisonte generaly for euere; & alle signes fro the heued of
libra vn-to þe ende of pisces ben cleped signes of the sowth fro   [Fol. 21]
the Equinoxial;   ¶ & thise signes arisen euer-mo by-twyxe the
verrey est & the verrey sowth in owre orisonte.   ¶ Also euery signe
16 by-twixe the heued of capricorne vn-to the ende of geminis ariseth
on owre Orisonte in lasse than 2 howres equales; ¶ & thise same
signes, fro the heued of capricorne v̄n-to the ende of geminis, ben
cleped tortuos signes or kroked signes / for they arisen embelif on
20 oure Orisonte;   ¶ & thise crokede signes ben obedient to the signes
þat ben of riht Assencioun.   ¶ The signes of riht assencioun ben
fro the heued of cancer to þe [ende] of sagittare;   ¶ & thise signes
arisen more vpriht, & they ben called eke souereyn signes; ¶ &
24 euerich of hem ariseth in mor space than in to howres.   ¶ Of which

---

labe, on as large a scale as is convenient. It is done by observing where
the ends of the given sign, estimated along the *outer* rim of the zodiacal
circle in fig. 2, cross the *horizon obliquus* as the *rete* is turned about.
Thus, the beginning of Aries lies on the *horizon obliquus*, and as the *rete*
revolves to the right, the end of it, on the outer rim, will at last lie exactly on the same curved line.   When this is the case, the *rete* ought to
have moved through an angle of about 14°, as explained in § 26.   By
far the best way is to tabulate the results once for all, as I have there
done.   It is readily seen, from fig. 2, that the signs from Aries to Virgo
are *northern*, and from Libra to Pisces are *southern* signs. The signs from
Capricorn to Gemini are the *oblique* signs, or as Chaucer calls them,
"tortuous," and ascend in less than 2 hours; whilst the *direct* signs,
from Cancer to Sagittarius, take more than 2 hours to ascend; as shewn
in the table in § 26.   The *eastern* signs in fig. 2 are said to *obey to* the
corresponding *western* ones.

signes gemini obeieth to Cancer, ¶ & taurus to leo / Aries to virgo / pisces to libra, ¶ Aquarius to Scorpioun, and Capricorne to Sagittare. ¶ & thus euermo 2 signes that ben illike fer fro the heued of capricorne / obeien euerich of hem til other.   ¶ & for more declara- 28 cioun, lo here the figure.

29. To knowe Iustly the 4 quarters of the world, as est, west, north, & sowth.

[Ad cognoscendum euidenter quatuor partes mundi, scilicet, orientem, austrum, aquilonem, & occidentem.]

¶ Take the altitude of thy sonne whan the list / & note wel the quarter of the world in which the sonne is for the tyme by the Azymutz. ¶ turne thanne thin Astrolabie / & set the degree of the sonne in the Almykanteras of his altitude on thilke side þat the 4 sonne stant / as is the manere in takyng of howres ; ¶ & ley thy label on the degree of the sonne, And rikene how many degres of the bordure ben by-twixe the lyne Meridional & the point of thy label; & note wel þat nowmbre.   ¶ * Turne thanne a-gayn [* Fol. 21 b.] 8 thyn Astralabie, & set the point of thy gret Rewle ther thow takest thyne Altitudes / vp-on as many degrees in his bordure fro his Meridional as was the point of thy label fro the lyne Meridional on the wombe-side. ¶ tak thanne thyn Astrolabie with bothe handes sadly 12 & slely, & lat the sonne shyne thorow bothe holes of thy rewle ; ¶ & sleyly in thilke shynynge lat thyn Astrelabie kowch adown euene vp-on a smothe grond, & thanne wol the verrey lyne Merydional of thyn

---

29. Here *both* sides of the Astrolabe are used, the "rewle" being made to revolve at the *back*, and the "label" in *front*, as usual. First, by the back of the instrument and the "rewle," take the sun's altitude. Turn the Astrolabe round, and set the sun's degree at the right altitude among the almicanteras, and then observe, by help of the label, how far the sun is from the meridian. Again turn the instrument ·round, and set the "rewle" as far from the meridian as the label was. Then, holding the instrument as near the ground and as horizontal as possible, let the sun shine through the holes of the "rewle," and immediately after lay the Astrolabe down, without altering the azimuthal direction of the meridional line. It is clear that this line will then point southwards, and the other points of the compass will also be known.

16 Astrolabie lye euene sowth, & the est lyne wole lie est, & the west
lyne west, & north lyne north, so þat thow werke softly & avisely in
the cowchyng ; & thus hastow the 4 quarters of the firmament.
¶ & for the more declaracioun, lo here the figure.

30. To knowe the Altitude of planetes fro the wey of the sonne, whether so they be north or sowth fro the forseide wey.

[Ad cognoscendum altitudinem planetarum a cursu solis, utrum
sint in parte australi vel boreali a cursu supra dicto.]

¶ Lok whan þat a planete is in the lyne Meridional, yif þat hir
altitude be of the same heyhte þat is the degree of the sonne for þat
day, & than is the planete in the verrey wey of the sonne, ¶ & hath
4 no latitude. ¶ & yif the altitude of the planete be heyere than the
degree of the sonne, þan is the planete north fro the wey of the
sonne swych a quantite of latitude as [shewith] by thyn Almykanteras. & ʒif the altitude of the planete be lasse than the degree
8 [* Fol. 22] of the sonne / thanne * is the planete sowth fro the wey of
the sonne swich a quantite of latitude as [shewith] by thin almykanteras ; ¶ This is to seyn, fro the wey wher as the sonne wente thilke
day / but nat from the wey of the sonne in euery place of the zodiak.
12 ¶ & for the more declaracioun, lo here the figure.

31. To knowe the senyth of the arysing of the sonne, this is to seyn, the partie of the Orisonte in which þat the sonne arisith.

[Ad cognoscendum signum de ortu solis, scilicet, illam partem
orientis in qua oritur sol.]

¶ Thow most first considere þat the sonne ariseth nat al-wey
verrey est, but some tyme by north the est, and som tyme by sowthe

---

30. This turns upon the definition of the phrase "the wey of the
sonne." It does not mean the zodiacal circle, but the sun's apparent
path on a given day of the year. The sun's altitude changes but little
in one day, and is supposed here to remain the same throughout the time
that he is, on that day, visible. Thus, if the sun's altitude be $61\frac{1}{2}°$, the
way of the sun is a small circle, viz. the tropic of Cancer. If the planet
be then on the zodiac, in the 1st degree of Capricorn, it is 47° S. from
the way of the sun, and so on.

the est / Sothly the sonne ariseth neuer-mo verrey est in owere Orisonte, ¶ but he be in the heued of aries or libra. now is thin 4 Orisonte departed in 24 parties by thi azymutz, in significacion of 24 partiez of the world; al be it so þat shipmen rikne thilke partiez in 32 / thanne is ther no more but waite in which azymut þat thi sonne entreth at [h]is arisyng / & take ther the senyth of the arising of the 8 sonne. ¶ the manere of the deuisioun of thin Astralabie is this / I Mene as in this cas. ¶ First is it deuided in 4 plages principalx with the lyne þat goth from est to west, ¶ & than with a-nother lyne þat goth fro so[w]th to north. ¶ than is it deuided in smale partiez 12 of Azymutz, as est, and est by sowthe, whereas is the firste Azimut aboue the est lyne ; ¶ & so forth fro partie to partie / til þat thow come agayn vn-to the est lyne / thus maistow vnderstond also the senyth of any sterre, in which partie he riseth, &c. ¶ & for the more 16 declaracion, lo here the figure.

[Fol. 22 b.] **32. To knowe in which partie of the firmament is the coniunccioun.**

[Ad cognoscendum in qua parte firmamenti sunt coniuncciones solis & lune.]

¶ Considere the tyme of the coniuncc[i]on by thy kalender / as thus ; lok how many howres thilke coniunccion is fro the Midday of

---

31. The word "senyth" is here used in a peculiar sense ; it does not mean, as it should, the *zenith* point, or point directly overhead, but is made to imply the point on the horizon, (either falling upon an azimuthal line, or lying between two azimuths), which denotes the point of sunrise. In the Latin rubric, it is called *signum*. This point is found by actual observation of the sun at the time of rising. Chaucer's azimuths divide the horizon into 24 parts; but it is interesting to observe his remark, that "shipmen" divide the horizon into 32 parts, exactly as a compass is divided now-a-days. The reason for the division into 32 parts is obviously because this is the easiest way of reckoning the direction of the wind. For this purpose, the horizon is first divided into 4 parts ; each of these is halved, and each half-part is halved again. It is easy to observe if the wind lies half-way between S. and E., or half-way between S. and S.E., or again half-way between S. and S.S.E.; but the division into 24 parts would be unsuitable, because *third-parts* are much more difficult to estimate.

the day precedent, as shewith by the canoun of thi kalender ; ¶ rikne
4 thanne thilke nombre of howres in the bordure of thyn Astralabie /
as thow art wont to do in knowyng of the howres of the day or of
the nyht ; ¶ & ley thy label ouer the degree of the sonne ; ¶ &
thanne wol the point of thy label sitte vp-on the hour of the con-
8 iunccion. ¶ loke thanne in which Azymut the degree of thy sonne
sittith, & in that partie of the firmament is the coniunccioun. ¶ &
for the more declaracioun, lo here thy figure.

[Fol. 23] **33. To knowe the senyth of the Altitude of the sonne, &c.**

[Ad cognoscendum signa de altitudine solis.]

This is no mor to seyn but any tyme of the day tak the altitude
of the sonne, & by the Azymut in which he stondith, ¶ Maistou
sen in which partie of the firmament he is ¶ & [in] the same wyse
4 maistou sen, by the nyht, of any sterre, wheither the sterre sitte est
or west or north, or any partie by-twene, aftur the name of the
Azimut in which is the sterre. ¶ &· for the more declaracioun, lo
here the figure.

**34. To knowe sothly the degree of the longitude of the mone,
or of any planete þat hath no latitude for the tyme fro the
Ecliptik lyne.**

----

32. The Latin rubric interprets the conjunction to mean that of the
sun and moon. The time of this conjunction is to be ascertained from
a calendar. If, e.g. the calendar indicates 9 A.M. as the time of conjunction on the 12th day of March, when the sun is in the first point of
Aries, as in § **3**, the number of hours after the preceding midday is 21,
which answers to the letter X in the border (fig. 2). Turn the *rete* till
the first point of Aries lies under the label, which is made to point to X,
and the label shews at the same moment that the degree of the sun is
very nearly at the point where the equinoctial circle crosses the azimuthal circle which lies 50° to the E. of the meridian. Hence the conjunction takes place at a point of which the azimuth is 50° to the E. of
the S. point, or 5° to the eastward of the S.E. point. The proposition
merely amounts to finding the sun's azimuth at a given time.

33. Here "senyth" is again used to mean azimuth, and the pró-
position is, to find the sun's azimuth by taking his altitude, and setting
his degree at the right altitude on the almicanteras. Of course the two
co-ordinates, altitude and azimuth, readily indicate the sun's exact position ; and the same for any star or planet.

[Ad cognoscendum veraciter gradum de longitudine lune, vel alicuius planete qui non habet longitudinem pro tempore causante (*sic*) linea ecliptica.]

¶ Tak the altitude of the mone, & rikne thin altitude vp among thyne Almykanteras on which side that the Mone stande, & set there a prikke. ¶ Tak thenne anon riht, vp-on the mones side, the Altitude of any sterre fix which þat thow knowest, & set his Centre vp-on his **4** altitude Among thin Almykanteras ther the sterre is fownde. ¶ Waite thanne which degree of the zodiak [towchith] the prikke of the altitude of the mone, & tak ther the degree in which the mone standith. ¶ this conclusioun is verrey soth, yif the sterres in thin **8** Astrolabie stonden aftur * the trowthe ; of comune, tretis of [* Fol. 23 b.] Astralabie ne make non excepcioun wheyth[er] the mone haue latitude, or non / ne on wheither side of the mone the Altitude of the sterre fix be taken. ¶ And nota, þat yif the Mone [shewe] himself by **12** liht of day, than maistow wyrke this same conclusioun by the sonne, as wel as by the fix sterre. ¶ & for the more declaracioun, lo here thy figure.

35. This is the workinge of the conclusioun, to knowe yif þat any planete be directe or retrograde.

[Hec conclusio operatur ad cognoscendum si aliqua planeta sit directa uel retrograda.]

¶ Tak the altitude of any sterre þat is cleped a planete, ¶ & note it wel. ¶ & tak ek anon the altitude of any sterre fix that thow

---

34. The moon's latitude is never more than $5\frac{1}{4}°$ from the ecliptic, and this small distance is, " in common treatises of Astrolabie," altogether neglected ; so that it is supposed to move in the ecliptic. First, then, take the moon's altitude, say 30°. Next take the altitude of some bright star " on the moon's side," i. e. nearly in the same azimuth as the moon, taking care to choose a star which is represented upon the *Rete* by a pointed tongue. Bring this tongue's point to the right altitude among the almicanteras, and then see which degree of the ecliptic lies on the almicantera which denotes an altitude of 30°. This will give the moon's place, " if the stars in the Astrolabe be set after the truth," i. e. if the point of the tongue is exactly where it should be.

35. The motion of a planet is called *direct*, when it moves in the

knowest, & note it wel also. ¶ Come thanne agayn the thridde or
4 the ferthe nyht next folwyng ; for thanne shaltow aperceyue wel the
Moeuyng of a planete, wheither so he Moeue forthward or bakward.
¶ Awaite wel thanne w[h]an þat thi sterre fix is in the same altitude
þat she was w[h]an thow toke hir firste altitude ; ¶ and tak than eft-
8 sones the Altitude of the forseide planete, & note it wel. ¶ for trust
wel, yif so be þat the planete be on the riht side of the Meridional
lyne, so þat his seconde altitude be lasse than [h]is firste altitude was,
thanne is the planete directe. ¶ And yif he be on the west side in
12 that condicion / thanne is he retrograd. ¶ And yif so be þat this
[* Fol. 24] planete be vp-on the Est side whan [h]is altitude is * taken, so
þat his secounde altitude be more than [h]is firste altitude, thanne is
he retrograde, & yif he be on the west side, than is he directe. ¶ but
16 the contrarie of this parties is of the cours of the Moone ; for
[sothly] the Moone Moeuyth the contrarie from othere planetes as in
hire Episicle, but in non other manere. ¶ & for the more declaracioun, lo here thi figure.

36. The conclusiouns of equaciouns of howses, after the astrala-bie, &c.

[Conclusio de equacione domorum.]

Set the by-gynnyng of the degree þat assendith vp-on the ende of
the 8 howre inequal ; thanne wol the by-gynnyng of the 2 hows
sitte vp-on the lyne of Midnyht. ¶ rem[e]ue thanne the degree þat
4 assendith, & set him on the ende of the 10 howr inequal ; & thanne

---

direction of the succession of the zodiacal signs ; *retrograde*, when in
the contrary direction. When a planet is on the right or east side of
the Meridional line, and is moving forward along the signs, without increase of declination, its altitude will be less on the second occasion
than on the first at the moment when the altitude of the fixed star is the
same as before. The same is true if the planet be retrograde, and on
the western side. The contrary results occur when the second altitude
is greater than the first. But the great defect of this method is that it
may be rendered fallacious by a change in the planet's declination.

   36. See fig. 14, Plate VI. If the equinoctial circle in this figure be
supposed to be superposed upon that in fig. 5, Plate III, and be further
supposed to revolve backwards through an angle of about 60° till the point

wol the by-gynnyng of the 3 howis sitte vp-on the Midnyht lyne. ¶ bryng vp agayn the same degree þat assendith first / & set him vp-on the Orisonte / & thanne wol the be-gynnyng of the 4 howys sitte vp-on the lyne of Midnyht. ¶ tak thanne the nadir of the 8 degree þat first Assendith / & set him on the ende of the 2 howre inequal / & thanne wol the by-gynnyng of the 5 howys sitte vp-on the lyne of Midnyth ; set thanne the nadir of the assendent on the ende of the 4 howre, þan wol the bygynnyng of the 6 house sitte on 12 the Midnyht lyne. ¶ þe bygynnyng of the 7 hows is nadir of the Assendent / & the bygynnyng of the 8 hows is nadir of the 2 ; & þe by-gynnyng of the 9 hous is nadir of the 3 ; & þe by-gynnyng of þe 10 hows is the nadir of the 4 ; & þe bygynnyng of the 11 howys is 16 nader of the 5 ; & the bygynnyng of the 12 hows is nadir of the 6. ¶ & for the more declaracion, lo here the figure.

[Fol. 24 b.] ### 37. A-nother manere of equaciouns of howses by the Astrelabie.

[De aliqua forma equacionis domorum secundum astrolabium.]

¶ Tak thin assendent, & thanne hastow thi 4 Angles; for wel thow wost þat the opposit of thin assendent, þat is to seyn, thy by-gynnyng of the 7 howis, sit vp-on the west orizonte ; ¶ & the bygynnyng of the 10 howis sit vp-on the lyne Meridional ; ¶ & his 4 opposit vp-on the lyne of Mydnyht. ¶ Thanne ley thi label ouer

---

1 (fig. 14) rests upon the point where the 8th hour-line crosses the equinoc-tial, the beginning of the 2nd house will then be found to be on the line of midnight. Similarly, all the other results mentioned follow. For it is easily seen that each "house" occupies a space equal to 2 hours, so that the bringing of the 3rd house to the midnight line brings 1 to the 10th hour-line, and a similar placing of the 4th house brings 1 to the 12th hour-line, which is the *horizon obliquus* itself. Moving onward 2 more hours, the point 7 (the nadir of 1) comes to the end of the 2nd hour, whilst the 5th house comes to the north; and lastly, when 7 is at the end of the 4th hour, the 6th house is so placed. To find the nadir of a house, we have only to add 6 ; so that the 7th, 8th, 9th, 10th, 11th, and 12th houses are the nadirs of the 1st, 2nd, 3rd, 4th, 5th, and 6th houses respectively.

37. Again see fig. 14, Plate VI. Here the 10th house is at once seen to be on the meridional line. In the quadrant from 1 to 10, the

the degree þat assendet[h] / & rekne fro the point of thy label alle
the degrees in the bordure, til thow come to the Meridional lyne / &
8 departe alle thilke degrees in 3 euene parties, & take the euene
equacion of 3 ; for ley thy label ouer euerich of 3 parties, & [than]
maistow se by thy label in which degree of the zodiak [is] the by-
gynnyng of euerich of thise same howses fro the assendent / þat is to
12 seyn, the begynyng of the [12] howse nex[t] aboue thin assendent /
And [thanne] the begynnyng of the 11 howse, & thanne the 10 vp-
on the Meridional lyne / as I first seide. ¶ The same wyse wyrke
thow fro the assendent down to the lyne of Mydnyht / & thanne
16 thus hastow other 3 howses, þat is to seyn, the bygynnyng of the 2
& the 3 And the 4 howses ; thanne is [the] nader of thise 3 howsez
the by-gynnyng of the 3 howses þat folwen. ¶ & for the more declaracioun, lo here thi figure.

[Fol. 25] **38. To fynde the lyne Merydional to dwelle fix in any
certein place.**

[Ad inueniendum lineam meridionalem per subtiles operaciones.]

Tak a rond plate of metal, for [warpyng] the brodere the bettre ;
¶ & make ther-vpon [a] Iust compas, a lite with-in the bordure / &
ley this ronde plate vp-on an euene grond or on a[n] euene ston or on
4 a[n] euene stok fix in the gronde / & ley it euen bi a leuel ¶ & in centre
of the compas stike an euene pyn or a whir vp-riht / the smallere þe
betere ¶ set thy pyn by a plom-rewle euene vpryht ¶ & let this pyn
be no lengere than a quarter of the diametre of thi compas, fro the
8 centre. ¶ & waite bisily aboute 10 or 11 of the clokke, & whan the

---

even division of the quadrant into 3 parts shews the 12th and 11th houses.
Working downwards from 1, we get the 2nd and 3rd houses, and the
4th house beginning with the north line. The rest are easily found
from their nadirs.

38. This problem is discussed in arts. 144 and 145 of Hymers's Astronomy, 2nd ed. 1840, p. 84. The words " for warpyng " mean " to
prevent the errors which may arise from the plate becoming warped."
The " broader " of course means " the larger." See fig. 15, Plate VI.
If the shadow of the sun be observed at a time *before* midday when its
extremity just enters within the circle, and again at a time *after* midday

sonne shynyth, whan the shadwe of the pyn entreth any-thyng
with-in the cercle of thi plate an her-mele, ¶ & mark ther a prikke
with inke.   Abide thanne stille waityng on the sonne aftur 1 of the
clokke, til that the schadwe of the wyr or of the pyn passe ony-thyng 12
owt of the cercle of the compas, be it neuer so lite / & set ther
a-nother prikke of ynke.   ¶ take than a compas, and mesure euene
the Middel by-twixe bothe prikkes, & set þer a prikke,   ¶ take
thanne a rewle / & draw a strike, euene alyne fro the pyn vn-to the 16
Middel prikke ;   ¶ & tak ther thy lyne Meridional for euere-mo, as in
that same place.   ¶ & yif thow drawe a cros-lyne ouer-thwart the
compas Iustly ouer the lyne Meridional, than hastow est and west
& sowth / &, par consequence, than the nader of the sowth lyne is 20
the north lyne.   ¶ & for more declaracioun, lo here thi figure.

[Fol. 25 b.] **39. Descripcion of the Meridional lyne, of longitudes, &
latitudes of Citees and townes from on to a-no[t]her of clymatz.**

This lyne Meridional ys but a Maner descripcion [of lyne]
ymagined, that passeth vpon the pooles of þis world And by the
cenyth of owre heued / And hit is [ycleped the] lyne Meridional / for
in what place þat any maner man [ys at] any tyme of the yer / whan 4
that the sonne [by moeuyng] of the firmament cometh to his verrey
[meridian] place / than is hit verrey Midday, þat we clepen owre
noon, ¶ As to thilke man ;   ¶ And therfore ys it clepid þe lyne of

---

when it is just passing beyond the circle, the altitude of the sun at these
two observations must be the same, and the south-line must lie half-way
between the two shadows.   In the figure, S and S' are the 2 positions of
the sun, OT the rod, Ot and Ot' the shadows, and OR the direction of the
south line.   Ott' is the metal disc.

39. This begins with an explanation of the terms "meridian" and
"longitude."   "They chaungen here Almikanteras" means that they
differ in latitude.   But, when Chaucer speaks of the longitude and
latitude of a "climate," he means the length and breadth of it.   A
"climate" (clima) is a belt of the earth included between two fixed
parallels of latitude.   The ancients reckoned seven climates ; in the sixteenth century there were nine.   The "latitude of the climate" is the
breadth of this belt ; the "longitude" of it he seems to consider as measured along lines lying equidistant between the parallels of latitude of

8 Midday. ¶ And no*ta*, for euermo, of [2 citees] or of 2 Townes, of whiche þat o town aprochith [more] towarde the Est þan doth þat other town, ¶ Truste wel that thylke townes han diuerse Meridians.

¶ Nota also, that the Arch of the Equinoxial that is [conteyned] or 12 bownded by-twyxe the 2 Meridians ys cleped þe longitude of the town. ¶ And [yf] so be þat two townes haue illike Meridian, or on Meridian, ¶ Than is the distance of hem bothe ylike fer fro the Est / & the *contra*rie. And in this Manere they chaunge nat her Meridian, 16 ¶ But sothly they chau*n*gen here Almikanteras, For the enhausyng of the pool and the distance of the sonne. ¶ The longitude of a clymat ys a lyne ymagined fro Est to west, illike distant by-twene · them alle. ¶ þe latitude of a climat is a lyne ymagined from north 20 [* Fol. 26] to south þe space of the erthe, fro the byginnyng *of the firste clymat vnto to the verrey ende of the [same] climat, euene directe agayns [þe poole Artik.] ¶ Thus seyn some Auctours / And som*m*e of hem seyn þat yif men clepen þe latitude, thay mene the arch meri- 24 dian þat is contiened or [inter]cept by-twixe the cenyth and the equinoxial. Thanne sey þey that the distaunce fro the equinoxial vnto þe ende of a clymat, euene agayns þe pool artyk, ys the latitude of a climat for sothe. ¶ & for more declaraciou*n*, lo here thi *figure*.

---

the places from which the climates are named. See Stöffler, fol. 20 *b*. ; and Petri Apiani Cosmographia, per Gemmam Phrysium restituta, ed. 1574, fol. 7 *b*. The seven climates were as follows :—

1. That whose central line passes through Meroë (lat. 17°) ; from nearly 13° to nearly 20°.
2. Central line, through Syene (lat. 24°) ; from 20° to 27°, nearly.
3. Central line through Alexandria (lat. 31°) ; from 27° to 34°, nearly.
4. Central line through Rhodes (lat. 36°) ; from 34° to 39°, nearly.
5. Central line through Rome (lat. 41°) ; from 39° to 43°, nearly.
6. Central line through Borysthenes (lat. 45°) ; from 43° to 47°.
7. Through the Riphæan mountains (lat. 48°) ; from 47° to 50°. But Chaucer must have included an *eighth* climate (called *ultra Mæotides paludes*) from 50° to 56° ; and a *ninth*, from 56° to the pole. The part of the earth to the north of the 7th climate was considered by the ancients to be uninhabitable. A rough drawing of these climates is given in MS. Camb. Univ. Lib. Ii. 3. 3, fol. 33 *b*

40. To knowe with which degree of the zodiak þat any planete Assendith on the Orisonte, wheyther so that his latitude be north or sowth.

¶ Knowe be thin almenak the degree of the Ecliptik of any signe in which þat the planete is rekned for to be / & that is cleped the degree of his longitude ; & knowe Also the degree of his lati- tude fro the Ecliptik, north or sowth. ¶ And by this samples 4 folwynge in special / maistow wyrke for [sothe] in euery signe of the zodiak. the degree of [longitude] par auenture, of venus or of another planete was 6 of Capricorne, & the latitude of him was northward 2 degrees fro the Ecliptik lyne. I tok a subtil compas, 8 & cleped þat on poynt of my compas A, & þat other poynt F. ¶ Than [tok] I the point of A, & set it in [the] Ecliptik line euene in my zodiak, in the degree of the longitude of venus / þat is to seyn, in the 6 degree of Capricorne ¶ & thanne set I the point of F vpward 12 in the same *signe, bycause þat the latitude was north, vp-on [* Fol. 26 b.] the latitude of venus, that is to seyn, in the 6 degree fro the heued of capricorne ; & thus hauy 2 degrees by-twixe my to prikkes ; than leide I down softely my compas, ¶ & sette the degree of the longi- 16 tude vp-on the Orisonte / tho tok I & wexede my label in Maner of a peyre tables to resceyue distynctly the prikkes of my compas. ¶ Tho tok I this forseide label, & leide it fix ouer the degree of my longitude / tho tok I vp my compas, ¶ & sette the point of A in the 20 wex on my label, as euene as y kowde gesse ouer the Ecliptik lyne, in the ende of [the] longitude / & sette the point of F endlang in my label vp-on the space of the latitude, inwarde & ouer the zodiak, that

---

40. The longitude and latitude of a planet being ascertained from an almanac, we can find with what degree it ascends. For example, given that the longitude of Venus is 6° of Capricorn, and her N. latitude 2°. Set the one leg of a compass upon the degree of longitude, and extend the other till the distance between the two legs is 2° of latitude, from that point inward, i. e. northward. The 6th degree of Capricorn is now to be set on the horizon, the label (slightly coated with wax) to be made to point to the same degree, and the north latitude is set off upon the

24 is to seyn, north-ward fro the Ecliptik // than leide I down my *com-*
pas & lokede wel in the wey vpon the prikke of A & of F ; tho
turned I my Riet til þat the prikke of F sat vp-on the Orisonte /
than saw I wel þat the body of venus, in hir latitude of [2] degrees
28 septe*n*trionalis, assendid, in the ende of the 6 degree, in the heued
of capricorne.   ¶ And no*t*a, þat in the same man*er* maistow wyrke
wit*h* any latitude septe*n*trional [in alle] signes ; but sothly the
latitude M*e*ridional of a planete in Capricorne may not be take, by
32 cause of the litel space by-twixe the Ecliptik / & the bordure of the
Astrelabie ; but sothly, in alle other [signes] it May.

[Fol. 27] ¶ Also the degree, pa*r* auent*u*re, of Iuppit*er* or of a-nother
planete, was in the furst degree of pisc*es* in longitude / & his lati-
36 tude was 3 degrees M*e*ridional ; tho tok I the point of A / & set it
in the firste degree of pisc*es* on the Ecliptik / & thanne set I the
point of F downward in the same signe, by cause þat the latitude was
sowth 3 degrees / þat is to seyñ, fro the heued of pisces / & thus hauy
40 3 degrees by-twixe bothe prikkes ; thanne sette I the degree of the
longitude vp-on the Orisonte ; tho tok I my label / & leide it fix vp-
on the degree of the longitude ; tho sette I the point of A on my
label, euene ou*er* the Ecliptik lyne in the ende euene of the degree of
44 the longitude / & set the point of [F endlang] in my label the space
[of] 3 [degrees] of the latitude fro the zodiak, this is to seyn,
sowthward fro the Ecliptik, toward the bordure ; and turned my

---

wax by help of the compass.   The spot thus marking the planet's position
is, by a very slight movement of the *Rete*, to be brought upon the horizon, and it will be found that the planet (situated 2° N. of the 6th degree) ascends together with the *head* (or beginning of the sign) of Capricorn.   This result, which is not *quite* exact, is easily tested by a globe.
When the latitude of the planet is *south*, its place cannot well be found
when in Capricorn, for want of space at the edge of the Astrolabe.

As a second example, it will be found that, when Jupiter's longitude
is at the *end* of 1° of Pisces, and his latitude 3° south, he ascends together
with the 14th of Pisces, nearly.   This is easily verified by a globe, which
solves all such problems very readily.

It is a singular fact that most of the best MSS. leave off at the word
"howre," leaving the last sentence incomplete.   For the last five words
—"þou shalt do wel ynow"—which I quote from the MS. in St. John's
College, Cambridge, see p. 52.

Riet [til] the prikke of F sat vp-on the Orisonte; thanne [saw] I wel
þat the body of Iuppiter, in his latitude of 3 degrees Meridional, **48**
ascendit with 14 degrees of pisces *in horoscopo* / &. in this Maner
maistow wyrke with any latitude Meridional, as I first seide, saue in
Capricorne / And yif thow wolt pleie this craft with þe arisyng of
the Mone, loke thow rekne wel her cours howre by howre; for she ne **52**
dwellith nat in a degree of [hire] longitude but [a] litel while, as thow
wel knowest / but natheles, yif thow rekne hir verreye Moeuyng by
thy tables howre after howre—

**Explicit tractatus de Conclusionibus Astrolabii compilatus per
Galfridum Chauciers ad Filium suum Lodewicum scolarem tunc
temporis Oxonie ac sub tutela illius nobilissimi Philosophi
Magistri N. Strode, &c.**

—[þou shalt do wel ynow.

*   *   *   *

41. Vmbra Recta.

[* Fol. 32] Ʒif it so be þat þou wylt werke be *vmbra* *recta, & þou may come to þe bas of þe towr*e*, in þis man*er* þou schalt werke.   Tak þe altitude of þe tou*r* be boþe holes, so þat þi rewle ligge euyn in a 4 poynt.   Ensample as þus : y see hym þorw at þe poynt of 4 ; þan mete y þe space be-twen me & þe tou*r*, & y fynde yt 20 feet ; þan be-holde y how 4 ys to 12, riʒt so is the space be[-twen] þe & þe tou*r* to þe altitude of þe tou*r*.   [For] 4 is þe [þ]ridde p*art* of 12, so is þe 8 space be-twen þe & þe tou*r* þe þridde p*art* of þe altitude of þe tou*r* ; þan þries 20 feet ys þe heyʒte of þe *tour*, wiþ addyng of þyn owne p*er*sone to þyn eye ; & þis rewle is so general in *vmbra recta*, fro þe poyn[t] of oon to 12.   And ʒif þi rewle falle vppon 5, þan is 5 12 12-p*artyes* of þe heyʒt þe space be-twen þe & þe towre ; wyþ addyng of þyn owne heyʒth.

42. Vmbra Versa.

Anoþ*er* man*er* of werkynge, be *vmbra versa*.   Ʒif so be þat þou may nat come to þe bas of þe tou*r*, y [see] hym þorw þe no*m*bre of 1 ; y [* Fol. 32 b.] sette þ*er* a prikke *at my foote ; þan goo [y] ner to þe tou*r*, 4 & y see hym þorw at þe poynt of 2, & þere y sette a-noþ*er* prikke ; &

---

41.   Sections 41—43 and 41*a*—42*b* are from the MS. in St John's College, Cambridge.   For the scale of *umbra recta*, see fig. 1, Plate I. Observe that the *umbra recta* is used where the angle of elevation of an object is greater than 45° ; the *umbra versa*, where it is less.   See also fig. 16, Plate VI ; where, if AC be the height of the tower, BC the same height *minus* the height of the observer's eye (supposed to be placed at E), and EB the distance of the observer from the tower, then *bc* : E*b* : : EB : BC.   But E*b* is reckoned as 12, and if *bc* be 4, we find that BC is 3 EB, i. e. 60 feet, when EB is 20.   Hence AC is 60 feet, *plus* the height of the observer's eye.   The last sentence is to be read thus—"And if thy 'rewle' fall upon 5, then are 5-12ths of the height equivalent to the space between thee and the tower (with addition of thine own height)."   The MS. reads "5 12-p*artyes* þe heyʒt *of* þe space," &c. ; but the word *of* must be transposed, in order to make sense.   It is clear that, if *bc* = 5, then 5 : 12 : : EB : BC, which is the same as saying that EB = $\frac{5}{12}$ BC.   Conversely, BC is $\frac{12}{5}$ EB = 48, if EB = 20.

42. See fig. 1, Plate I.   See also fig. 17, Plate VI.   Let E*b* = 12,

y be-holde how 1 hath hy*m* to 12, & *þer* fynde y *þat* yt hath hym
twelfe sithes ; *þan* be-holde y how 2 hath hy*m* to 12, & *þou* schalt
fynde it sexe sy*þ*es ; *þan þou* schalt fynde *þat* [as] 12 [above] 6 [is
þe] numbre of 6, Ry*3*t so is *þe* space be-twen *þi* too prikkis *þe* space **8**
of 6 tymes *þyn* altitude. & note, *þat* at *þe* ferste altitude of 1, *þou*
settest a prikke, & aftyrward, wha*n þou* [seest] hy*m* at 2, *þer þou*
settest an-o*þer* prikke, *þan þou* fyndest betwen too prikkys [60]
fett ; *þan þou* schalt fynde *þat* [10 is *þe* 6-party of 60.   And *þe*n is **12**
10 fete] *þe* altitude of *þe* tou*r*.   [For] o*þer* poyntis, *3*if yt fylle in
vmbra versa, as *þ*us : y sette caas it fill vppon [2], & at *þe* secunde
vppon [3] ; *þan* schalt *þou* fynde *þat* [2] is [6] *partyes* of 12 ; [and 3
is 4 *partyes* of 12] ; *þan* passe*þ* 6 4, be nombre of 2 ; so ys *þe* space **16**
be[twen] too prikkes twyes *þe* hey*3*te of *þe* tou*r*. & *3*if *þe* differens were
*þ*ries, *þan* schulde it be [*þ*re] tymes ; & *þ*us mayst *þou* werke fro 2 to
12 ; & *3*if yt *be 4, 4 tymes ; or 5, 5 tymes, & *sic de ceteris.*   [* Fol. 33]

43. Vmbra [Recta].

An o*þer* mane*r* of wyrkyng be *vmbra* [*recta*].   *3*if it so be *þat þou*
mayst nat come to *þe* baas of *þe* tou*r*, yn *þ*is mane*r þou* schalt werke.
Sette *þi* Rewle vppon [1] till *þou* see *þe* altitude, & sette at *þi* foot a
prikke. *þan* sette *þi* Rewle vppon 2, & be-holde what ys *þe* diff[e]renso **4**
be-twen 1 and 2, & *þou* shalt fynde *þat* it is 1. *þan* mete *þe* space
be-twe*n* too prikkes, & *þat* ys *þe* 12 *partie* of *þe* altitude of *þe* tou*r* ;
& *3*if *þer* were 2, yt were *þe* 6 *partye* ; & *3*if *þer* were 3, *þe* 4 *partye*,
& *sic deinceps.*   And note, *3*if it were 5, yt were *þe* 5 party of 12 ; **8·**
& 7, 7 *party* of 12 ; and note, at *þe* altitude of *þi* conclusiou*n*, adde
þe stature of *þyn* heythe.

*   *   *   *

---

bc = 1 ; also E′*b*′ = 12, *b*′*c*′ = 2 ; then EB = 12 BC, E′B = 6 BC ;
therefore EE′ = 6 BC.   If EE′ = 60 feet, then BC = $\frac{1}{6}$ EE′ = 10 feet.
To get the whole height, add the height of the eye.   The last part of the
article, beginning "For o*þer* poyntis," is altogether corrupt in the MS.

43. Here *versa* (in the MS.) is certainly miswritten for *recta*.   See
fig. 18, Plate VI.   Here E*b* = E′*b*′ = 12 ; *b*′*c*′ = 1, *bc* = 2.   Hence
E′B = $\frac{1}{12}$ BC, EB = $\frac{2}{12}$ BC, whence EE′ = $\frac{1}{12}$ BC.   Or again, if *bc*
become = 3, 4, 5, &c., successively, whilst *b*′*c*′ remains = 1, then EE′
is successively = $\frac{2}{12}$ or $\frac{1}{6}$, $\frac{3}{12}$ or $\frac{1}{4}$, $\frac{4}{12}$, &c.   Afterwards, add in the
height of E.

44. Another maner conclusion, to knowe the mene mote & þe argumentis of any planete. To know the mene mote & the argumentis of euery planete fro ȝere to ȝere, from day to day, from owre to owre, And from smale fraccionis infinite.

[**Ad cognoscendum medios motus & argumenta de hora in horam cuiuslibet planete, de anno in annum, de die in diem.**]

[* Fol. 106] In this maner shalt þou worche: *consider* thy rote furst, *the wyche is made the begynning of the tabelis fro the ȝere of owre lord 1397, & entere hit in-to thy slate for the laste merydye of December ; 4 and þan *consider* þe ȝere of ovre lord, what is þe date, & be-hold wheþer thy date be more or lasse þan þe ȝere 1397. And yf hit so be þat hit be more, loke how many [ȝeris] hit passith, & with so many entere into thy tabelis in þe furst lyne þer as is wreten *anni collecti* 8 *& expansi.* And loke [where] the same planet is wreten in the hede of thy tabele, and than [loke] what þou findest in directe of the same ȝere of owre lord wyche is passid, be hit 8, or 9, or 10, [or what nombre þat euere it be, tyl þe tyme þat þou come to 20, or] 40, or 60. 12 And that þou fyndest in directe [wryte] in thy slate vnder thy rote, & adde hit [to-geder], and þat is thy mene mote, for the laste meridian of the december, for the same ȝere wyche þat þou [hast] purposid. And yf hit so be [þat] hit passe 20, *consider* welle **þat fro** 16 [1] to 20 ben *anni expansi*, And fro 20 to 3000 ben *anni collecti ;* and yf thy nombere passe 20, þan take þat þou findest in directe of 20, & yf hit be more, as 6 or 18, than take þat þou findist in directe there-of, that is to sayen, signes, degreis, Minutes, & secundis, and 20 adde [to-gedere] vn-to thy rote ; and thus to make rotes ; and note,

---

44. Sections 44 and 45 are from MS. Digby 72. This long explanation of the method of finding a planet's place depends upon the tables which were constructed for that purpose from observation. The general idea is this. The figures shewing a planet's position for the last day of December, 1397, give what is called the *root*, and afford us, in fact, a *starting-point* from which to measure. An "argument" is the angle upon which the tabulated quantity depends ; for example, a very important "argument" is the planet's *longitude*, upon which its *declination* may be made to depend, so as to admit of tabulation. The planet's declination for the given above-mentioned date being taken as the *root*, the planet's declination at a second date can be found from the

þat yf hit *so be [þat] the ȝere of ovre lord be [lasse] than [* Fol. 106 b.]
the rote, wyche is the ȝere of ovre lord 1397, than shalt þou write in
the same wyse furst thy rote in thy slate, and after entere in-to thy table
in the same ȝere [þat] be lasse, as I tauȝth be-fore; and þan consider how 24
many signes, degrees, Minutes, & secundis thyne entrynge conteynith.
And so be that [þer] be 2 entres, than adde hem togeder, & after
with-drawe hem from the rote, the ȝere of ovre lord 1397 ; and the
residue þat lewyth is thy mene mote fro the laste mer[y]die of 28
December, the wyche þou haste purposid ; and yf hit so be þat þou
wolt weten thy mene mote [for] eny day, or [for] ony fraccion of day,
in þis maner þou shalt worche. make thy rote fro the laste day of
Decembere in þe maner as I thaȝthe, and afterward behold how many 32
monythis, dayes, & howris ben passid from [þe] merydye of Decembere & with that entere [with þe] laste moneth þat is ful passid, and
take þat þou findest in directe of hym, & wryte hit in thy slate ; &
entere with as mony dayes as be more, and wryte þat þou findest in 36
directe of the same planete þat þou worchyst fore ; and in þe same
wyse in-to þe table of howris, for hovris þat ben passid, and adde
alle these to thy rote ; and the *residue is the mene mote for [* Fol. 107]
the same day & þe same hovre.   40

45. Another manere to knowe the mene mote.

Whan þou wolte make the mene mote of eny planete to be by
arsechieles tables, take thy rote, the wyche is for the ȝere of ovre
lord 1397 ; and yf so be that thy ȝere be passid the date, wryte that
date / and than write that nombere of the ȝeris. þan wyth-drawe þe 4
ȝeris oute of the ȝeris that ben passid that rote. Ensampulle as thus :
the ȝere of ovre lord 1400, I-wryton precise, my rote ; þan wrote I

---

tables. If this second date be less than 20 years afterwards, the increase
of motion is set down separately for each year, viz. so much in 1 year, so
much in 2 years, and so on.   These separate years are called *anni expansi*.
But when the increase during a large round number of years (such as
20, 40, or 60 years at once) is allowed for, such years are called *anni
collecti*.   For example, a period of 27 years includes 20 years *taken together*, and 7 separate or *expanse* years.   The mean motion during smaller
periods of time, such as months, days, and hours, is added on afterwards.
45. Here the author enters a little more into particulars. If the mean

furst 1400. And vnder that nombere I wrote a 1397 ; þan wiþ-
8 drowe I the laste nombere owtc of þat, and þan fond I þe residue was
3 ȝere ; I wyst þat 3 ȝere was passid fro the rote, þe wyche was
wreten in my tabelis. Than after-ward soȝth I in my tabelis þe *annis*
collectis & expansis, & amonge myne expanse ȝeris fond I 3 ȝere. þan
12 toke I alle þe signes, degreis, & minutes, þat I fond directe vnder þe
same planete þat I wroȝth fore, & wrote so many signes, degreis, &
Minutes in my slate, & after-ward added I too signes, degreis,
Minutes, & Secundis, þe wiche I fond in my rote the ȝere of owre
16 lord 1397 ; And kepte the residue ; & þan had I the mene mote for
þe laste day of Decembere. And yf þou woldest wete þe mene mote of
any planete in March, Aprile, or may, oþer in any oþer tyme or monyth
[* Fol. 107 b.] of the ȝere, loke how many monethes & dayes *ben passid
20 from þe laste day of Decembere, the ȝere of owre lord 1400 ; and soe
wiþ monithis & dayes entere in-to þy table þer þou findist thy mene
mote I-wreten in monethis & dayes, and take alle þe signes, degrees,
Minutes, & secundis þat þou findest I-wrete in directe of thy monethis,
24 and [adde] to signes, degreis, Minutes, & secundis þat þou findest wiþ
thy rote þe ȝere of ovre lord 1400, and the residue þat leuyth is þe
mene mote for that same day. And note yf hit so be that þou woldest
[wete þe] mene mote in ony ȝere þat is lasse þan thy rote, wiþ-drawe
28 þe nombere of so many ȝeris as hit is lasse þan þe ȝere of ovre lord a
1397, & kepe þe residue ; & so many ȝeris, monythis, & dayes entere
in-to thy tabelis of thy mene mote. And take alle the signes,
degreis, and Minutes, [and] Secundis, þat þou findest in directe of alle
32 þe ȝeris, monythis, & dayes, & wryte hem in þy slate ; and abowe
þilke nombere write þe signes, degreis, Minutes, & secundis, þe wyche
þou findest wiþ thy rote þe ȝere of ovre lord a 1397 ; & wiþ-drawe
alle þe nethere signis & degreis fro þe signes & degrees, Minutes, &
36 Secundis of oþer signes wiþ thy rote, and thy residue þat lewyth is
thy mene mote for þat day.

---

motion be required for the year 1400, 3 years later than the starting-
point, look for 3 in the table of expanse years, and add the result to the
number already corresponding to the " root," which is calculated for the
last day of December, 1397. Allow for months and days afterwards. For
a date earlier than 1397 the process is just reversed, involving subtrac-
tion instead of addition.

*   *   *   *

41*a*. Vmbra Recta.

[Fol. 34 *b*.] Ȝif þi rewle falle vppon þe 8 poynt on riȝt schadwe, þan make þi figure of 8 ; þan loke how moche space of feet ys be-twen þe & þe tour, & multiplie þat be 12, & whan þou [hast] multiplied it, þan diuide yt be þe same nombre of 8, & kepe þe residue, & adde þerto 4 vp to þyn eye to þe residue, & þat schal be þe verry heyȝt of þe tour. & þus mayst þou werke on þe same wyse, fro 1 to 12.

41*b*. Vmbra Recta.

An-oþer mauer of werkyng vppon þe same syde. Loke vppon whych poynt þi Rewle falliþ whan þou seest þe top of þe tour þorow too litil holes, & mete þan þe space fro þi foot to þe baas of þe tour ; & ryȝt *as the nombre of þy poynt hath hym-self to 12, ryȝt so [* Fol. 36] 4 þe mesure be-twen þe & þe tour haþ hym-self to þe heiȝte of þe same tour. Ensample : y sette caas þi rewle falle vpon 8, þan ys 8 to-þridd partyes of 12 ; so þe space ys þe too-þridd partyes of þe tour.

42*a*. Vmbra Versa.

To knowe þe [heyth by þy] poyntes of *vmbra versa*. Ȝif þy rewle falle vppon 3, whan þou seest þe top of þe tour / sett a prikke þere-as þi foot stont ; & goo ner tyl þou mayst see þe same top at þe poynt of 4, & sette þer anoþer lyk prikke / þan mete how many foot 4 ben be-twen þe too prikkis, & adde þe lengþe vp to þyn eye þer-to ; & þat schal be þe heyȝte of þe tour. And note, þat 3 ys fourþe party of 12, & 4 is þe þridde party of 12. Now passeþ 4 þe nombre of 3 be þe distaunce of 1 ; þerfore þe same space, wyþ þyn heyȝt to 8

41*a*. This comes to precisely the same as Art. **41**, but is expressed with a slight difference. See fig. 16, where, if $bc = 8$, then $BC = \frac{12}{8}$ EB.

41*b*. Merely another repetition of Art. **41**. It is hard to see why it should be thus repeated in almost the same words. If $bc = 8$ in fig. 16, then $EB = \frac{8}{12}$ BC $= \frac{2}{3}$ BC. The only difference is that it inverts the equation in the last article.

42*a*. This is only a particular case of Art. **42**. If we can get $bc = 3$, and $b'c' = 4$, the equations become $EB = 4$ BC, $E'B = 3$ BC ; whence $EE' = $ BC, a very convenient result. See fig. 17.

þyn eye, ys þe heyȝt of þe *tour*. & ȝif it so be þat þe*r* be 2 or 3 distaunce in þe nombres, so schulde þe mesures be-twen þe prikkes be
twyes or þries þe heyȝte of þe *tour*.

43*a*. Ad cognoscendum altitudinem alicuius rei per vmbram [rectam].

[Fol. 36 *b*.] To knowe þe heyȝte of þynges, ȝif þou mayst [nat] come
to þe bas of a þyng. sette þy rewle vppon what þou wylt, so þat
þou may see þe topp of þe þyng þorw þe too holes, & make a marke
4 þe*r* þy foot standeþ ; and goo neer or forþe*r* / till þou mayst see þorw
anoþe*r* poynt, & marke þe*r* a-noþe*r* marke ; & loke þan what ys þe
differense be-twen þe too poyntes in þe scale ; & riȝt as þat difference
haþ hym to 12, riȝt so þe space be-twen þe & þe too markys haþ hym
8 to þe heyȝte of þe þyng. Ensample : y set caas þou seest it þorw a
poynt of 4 ; aftyr, at þe poynt of 3. Now passiþ þe nombre of 4 þe
nombre of 3 be þe difference of 1, and riȝt as þis difference 1 haþ
hym-self to 12, riȝt so þe mesure be-twen þe too markis haþ hy*m* to
12 þe heyȝte of þe þyng, puttyng to þe heyȝte of þi-self to þyn eye ; &
þus mayst þou werke fro 1 to 12.

42*b*. Per vmbram versam.

Furþermore, ȝif þou wilt knowe in *vmbra versa* / be þe craft of
vmbra recta, y suppose þou take þe altitude at þe poynt of 4, &
makest a marke, & þou goost neer tyl þou hast yt at þe poynt of 3,
4 [* Fol. 37] & þan makyst *þou þe*r* an-noþe*r* mark. þan muste þou deuide
144 be eche of þe poyntes be-fornseyd, [as] þus : ȝif þou deuide 144

43*a*. The reading *versam* (as in the MS.) is absurd. We must also
read " *nat* come," as, if the base were approachable, no such trouble need
be taken ; see Art. 41. In fact, the present article is a mere repetition of
Art. 43, with different numbers, and with a slight difference in the
method of expressing the result. In fig. 18, if $b'c' = 3$, $bc = 4$, we have
$E'B = \frac{3}{12} BC$, $EB = \frac{4}{12} BC$; or, subtracting, $EE' = \frac{4-3}{12} BC$; or BC
$= 12 EE'$. Then add the height of E, viz. E*a*, which $= AB$.

42*b*. Here, " by the craft of *Umbra Recta*" signifies, by a method
similar to that in the last article. In fig. 17, if $bc = 3$, $b'c' = 4$, then
$EB = \frac{12}{3} BC$, and $E'B = \frac{12}{4} BC$. Hence $EE' = (\frac{12}{3} - \frac{12}{4}) BC$. This
may be written, $EE' = (\frac{144}{3} - \frac{144}{4}) \frac{BC}{12}$, or $EE' : BC : : \frac{144}{3} - \frac{144}{4} : 12$;

be [4,] & þe nombre þat comeþ þer-of schal þe 36, & ȝif þou deuide 144 be 3, & þe nombre þat comeþ þer-of schal be 48, þanne loke what ys þe difference be-twen 36 & 48, & þer schalt þou fynde 12 ; **8** and ryȝt as 12 haþ hym to 12, ryȝt so þe space be-twen too prikkes haþ hym to þe altitude of þe þyng.

*   *   *   *

46. For to knowe at what houre of þe day, or of the night, shal be Flode or ebbe.

First wite thou certeinly, how that haven stondith, that thou list to werke fore ; þat is to say in whiche place of the firmament the mone beyng, makiþ fulle see. Than awayte þou redily in what degree of þe zodiak þat þe mone at þat tyme is ynne. Bringe furth **4** than þe labelle, & sett the point therof in þat same cost þat the mone makiþ flode, and sett þou þere þe degre of þe mone according wiþ þe egge of þe label. Than afterward awayte where is than þe degre of the sonne, at þat tyme. Remeue þou than þe label fro the mone, & **8** bringe & sette [it] iustly vpon þe degre of þe sonne. And þe point of þe label shal þan declare to þe, at what houre of þe day or of þe night shal be flode. And þere also maist þou wite by þe same point of þe label, wheþir it be, at þat same, flode or ebbe, or half flode, or **12** quarter flode, or ebbe, or half or quarter ebbe ; or ellis at what houre it was last, or shalbe next by night or by day, þou þan [maist] esely

---

or : : 12 : 12 ; whence EE′ = BC. This is nothing but Art. **42** in a rather clumsier shape.

Hence it appears that there are here but 3 independent propositions, viz. those in articles **41**, **42**, and **43**, corresponding to figs. 16, 17, and 18 respectively. Arts. **41**a and **41**b are mere repetitions of **41** ; **43**a of **43** ; and **42**a and **42**b, of **42**.

46. This article is probably not Chaucer's. It is found in MS. Bodley 619, and perhaps nowhere else. What it asserts comes to this. Suppose it be noted, that at a given place, there is a full flood when the moon is in a certain quarter ; say, e. g. when the moon is due east. And suppose that, at the time of observation, the moon's actual longitude is such that it is in the first point of Cancer. Make the label point due east ; then bring the first point of Cancer to the east by turning the *Rete* a quarter of the way round. Let the sun at the time be in the first point of Leo, and bring the label over this point by the motion of the

knowe, etc. Furþermore if it so be þat thou happe to worke for þis
16 matere aboute þe tyme of coniunccioun, bringe furþe þe degre of þe
mone wiþ þe labelle to þat coste as it is before seýde. but' than þou
shalt vnderstonde þat þou may not bringe furþe þe label fro þe degre
of þe mone as þou dide before ; For-why the sonne is þan in þe same
20 degre with the mone. And so þou may at þat tyme by þe point of
the labelle vnremevid knowe þe houre of þe flode or of þe ebbe as it
is before seyd, &c. And euermore as þou fїndest þe mone passe fro
þe sonne, so remeve þou þe labelle þan fro þe degre of þe mone, and
24 bringe it to the degre of þe sonne. And worke þou þan as þou dide
before, etc. Or ellis knowe þou what houre it is þat þou art inne, by
þin instrument. Than bringe þou furthe fro thennes þe labelle and
ley it vpon þe degre of þe mone, and þerby may þou wite also whan
28 it was flode, or whan it wol be next, be it nyght or day ; &c.

---

label only, keeping the *Rete* fixed. The label then points nearly to the
32nd degree near the letter Q, or about S.E. by E.; showing that the sun
is S.E. by E. (and the moon consequently due E.) at about 4 A.M. In
fact, the article merely asserts that the moon's place in the sky is known
from the sun's place, if the difference of their longitudes be known. At
the time of conjunction, the moon and sun are together, and the difference of their longitudes is zero, which much simplifies the problem. If
there is a flood tide when the moon is in the E., there is another when it
comes to the W., so that there is high water *twice* a day. It may be
doubted whether this proposition is of much practical utility.

CRITICAL NOTES.

TITLE. Tractatus, &c. ; adopted from the colophon. MS. F has "tractatus astrolabii." The other title, 'Bred and mylk for childeren,' is in MSS. B and E.

[The MSS. are as follows :—A. Cambridge Univ. Lib. Dd. 3. 53.—B. Bodley, E Museo 54.—C. Rawlinson 1370.—D. Ashmole 391.—E. Bodley 619.—F. Corpus 424.—G. Trin. Coll. Cam. R. 15. 18.—H. Sloane 314.—I. Sloane 261.—K. Rawlinson Misc. 3.—L. Addit. 23002. (B. M.)—M. St. John's Coll. Cam.—N. Digby 72.—O. Ashmole 360.—P. Camb. Univ. Lib. Dd. 12. 51.—Q. Ashmole 393. See the descriptions of them in the Preface.]

PROLOGUE. l. 26. thise B ; þese C ; *miswritten* this A ; see above, ll. 21, 22.

31. curious BC ; *miswritten* curios A.

36. nawht B ; nouȝt C ; *miswritten* nahwt A.

42. Astrologiens] *miswritten* Astrologens ABC ; but see l. 50 below.

48. practik B ; practyk CM ; *miswritten* practric A.

55. sonne BM ; *miswritten* som A. The seven words, *& tables—sonne*, are omitted in C.

57. a-nother B ; nother A ; oþer C.

58. clerks AB ; clerkus C ; *but the best spelling is* clerkes.

59. theorik BC ; thiorik A ; *but* A *has* theorik *in* l. 70.

60. þe C ; þᵉ B ; A *omits*.

PART I. § 1, l. 1 ; thowmbe B ; þombe CM ; *miswritten* towmbe A.
3. wol B ; wolde AC.

§ 2, l. 2. Astrelabie] *here miswritten* Asterlabie A.

N.B. *Rowm* is here an adjective, meaning *large, ample.* It is the right reading ; we find Rowm AB ; rowme C ; rvm M.

§ 3, l. 1. AB *omit* þe ; *in* C, *it is inserted in the margin ; in* M, *it is found in the text.*

2. resscyuyth B ; resseyueþ C ; receyueþ M ; *miswritten* resceiued A.

3. shapen B ; schapen CM ; *miswritten* spapen A.

4. declaracioun] *here written* decleracioun A. See sect. 4, l. 6.

§ 4, l. 5. remenant (see sect. 5, l. 5)] remenaunt C ; *miswritten* re menanañt A ; remonant B. downe BC ; doun M ; dowene A.

K

§ **5**, l. 2. lengþe C ; *miswritten* lenghthe A ; lenghte B.

§ **6**, l. 2. litul B ; lytel C ; lite A.

§ **7**, l. 6. by-twene BC ; by-thwene A.

8. that is] this B ; þis is [is *in* margin] C ; this, *altered to* that A. Perhaps the right reading is ' this is ' ; but it is immaterial.

§ **9**, l. 3. nombre AB ; noumbre C ; *but the sense requires the plural.*

§ **10**, l. 3. Septembre B] *miswritten* Sextembre A. 13. August BCM ; Augist A.

§ **12**, l. 5. The MSS. all[1] read—" *vmbra recta* or elles *vmbra extensa*, & the nether partie is cleped the *vmbra versa.*" This is wrong ; see the note on p. 7.

§ **13**, l. 2. a certein] *so in* AB ; CM *omit* a. But Chaucer certainly uses the phrase ' a certain ' ; cf. ' of unces a certain ', C. T. 16244 ; ' a certain of gold ', C. T. 16492.

§ **14**, l. 4. streyneþ C ; *miswritten* streynet AB.

N.B. The word *halt* for *holdeth*, and the expression *to hepe*, together, both occur in Troil. iii. 1770 :—

' And lost were al, that Love *halt* now *to hepe.*'

5. ymagyned C ; ymaginet B ; ymagynd A.

§ **15**, l. 2. w*ith* BC ; wit A.

§ **16**, l. 12. haue I C ; haue y M ; hauy B ; haue A.

§ **17**, l. 1. principal C ; tropikal AB ; M *om.* The reading *tropikal* is absurd, because there are but *two* such ; besides which, see l. 33 below.

6. ptholome] ptolomeys almagest M.

9. by-gynneth B ; bygynneþ C ; by-gynned A.

17. the nyht (*over an erasure*) B ; thee nyht (*over an erasure*) A ; þe niʒtes C ; þe nyʒtes M.

20. cleped C ; clopud A ; cheped B.

25. makeþ CM ; maked AB.

34. turnyth G ; turneþ C ; turned AB.

§ **19**, l. 3. ouerthwert M ; ouerþewart C ; ouertward A ; ouerthart B. *Read* ouerthwart ; see Ch. Knightes Tale, 1133.

8. for the more ; A *has* fore more *here ;* but see last lines of sections 17 and 18.

§ **20**, l. 1. azymutz C ; Azamutz B ; *miswritten* azymitz A ; *so in* l. 3, A *has the bad spelling* azimites.

4. figure ; here (*and sometimes elsewhere*) *miswritten* vigur A. Throughout the whole treatise, the scribe has commonly written " vigur" ; in many places, it has been corrected to " figure".

§ **21**, l. 14. the *supplied from* BC.

15. is B ; ys C ; *miswritten* his A.

26. where as C ; wher AB.

28. ymaginet AB ; ymagyned C ; *see* l. 25, *where, however, it is spelt* ymagened.

[1] As far as I can ascertain.

35. Minutes C; Minnutes B; Minites A.

39. swich B; sich C; swhich A.

43. understonde CM; vnderstonden AB.

53. ouerkeruyd A; ouerkerued B; ouerkerueth (*the latter part of the word over an erasure*) C; *see* l. 54.

56. here] *so elsewhere;* hir AB, *in this place.*

PART II. § **2**, l. 2. remewe CM; remue AB.

3. thorgh; *written* thorghw A; þorw M; þorwe C; to ʒow (!) B.

8. euer M; euere C; euery (*wrongly*) AB.

§ **3**, l. 9. sitten] *written* siten AB; sitte C; sittyn M.

11. owre] *written* howre AB; oure C.

22. down B; don A.

30, 31. A *has* 12 degres, *corrected to* 18 degres; B *has* 12 degrees; C *has* 18. The numbers in the MSS. in these propositions are some-what uncertain; it seems probable that some alteration was made by Chaucer himself.

The readings in MS. B give one set of calculations, which are no doubt the original ones; for in MS. A the same set is again found, but altered throughout, by the scribe who drew the diagrams. The sets of readings are these :—

Ll. 30, 31. 12 degrees B; *so in* A, *but altered to* 18; C *has* 18.

36. passed 9 of the clokke the space of 10 degrees B; *so in* A, *with* 9 *altered to* 8, *and* 10 *altered to* 2; C *has* ij *for* 9, *but agrees with* A *in the reading* 2.

38. fond ther 10 degrees of taurus B; *so in* A *originally, but* 10 *has been corrected to* 23, *and* libra *is written over an erasure.* C *agrees with neither, having* 20 *for* 10, *but agreeing with* A *as to* libra. The later MSS. sometimes vary from all these. See Chaucer's Astrolabe, ed. E. A. Brae, p. 34.

41. an *supplied from* C; AB *omit.*

§ **4**, 5. largest C; largesse AB.

6. vpon] upon C; *miswritten* vn AB.

7. forseide degre of his longitude] forseyde same degre of hys lon-gitude C; forseid same gre of his longitude P; forseyde latitude his longitude (*sic !*) AB.

8. planete ys C; *miswritten* planetes AB, *but* is *is added in margin of* A.

14. For " 25 degrees," all the MSS. have " 15 degrees." The mis-take is probably Chaucer's own; the correction was made by Mr Brae, who remarks that it is a mere translation from the Latin version of Ptolemy's Tetrabiblos, which has—" Signum ascendentis, quod est a *quinque* gradibus qui super horizontem ante ipsum ascenderant usque ad *viginti quinque* qui ad ascendentem remanserint"; Lib. iii. c. 10. In fact, it is clear that 25 must be added to 5 to make up the extent of a " house," which was 30 degrees.

15. ys like C; is lik P; *miswritten* illyk AB. in *is supplied from* GM; ABC *omit it.*

18. Astrologiens B ; Astrologens AC.

22. þe *supplied from* CP ; AB *omit.*

30. wel *supplied from* CPM ; AB *omit.*

34. than] þan CM ; þenne P ; AB *omit.*

37. *The number* 10 *is supplied from* C ; AB *omit.* It is obviously right, since the third part of 30 is 10.

39. some folk *supplied from* CPG ; AB *omit.*

41. yit is] AB *wrongly have* yit it is; *but* CPGM *omit it.*

§ 5, l. 3. by 2 & 2 ACG ; by 3 & 3 P ; *left blank in* B. Either reading makes sense, but it is clear that divisions representing three degrees each must have been very awkward.

10. of *supplied from* CPGM ; AB *omit.*

11. towcheth A ; toucheþ C ; towecheth B.

§ 6, l. 1. nadir B ; nadair AC.

5. est C ; west A (*which is absurd*) ; west (*corrected to* est) B.

9. sigue CGP ; signes ABM.

§ 7, l. 1. orisonte B ; *miswritten* oriensonte A.

§ 10, l. 3. than B ; þan C ; A *has* & by nyht, *which is absurd.*

4. A *omits* day with þe howr inequal of þe, *which is supplied from* BCP ; *the number* 30 *is also supplied from* BCM, *as* A *has a blank space there;* see l. 9.

6. answering] answerynge C ; answeryng P ; *miswritten* answerinc A ; answerit B.

§ 11, l. 8. by nyht B ; be nyhte A ; see l. 3 above, and l. 11 below.

12. *The number* 4 *is from* CP ; AB *omit.* þer *supplied from* PM ; þere C ; AB *omit.*

§ 12, l. 1. the *supplied from* BC ; A *omits.*

8. *The figure* 2 *is from* BCP ; G *has* secunde ; A *omits.*

10. entrith] entriþ P ; entryth G ; entreþ CM ; *miswritten* entrist AB.

§ 13, l. 4. this ys C ; this is G ; þis is M ; þat is P ; AB *omit* is.

§ 14, l. 9. The last line supplied from B.

§ 15, l. 5. varieth] varieþ CM ; varyiþ P ; variet ABG.

6. pointe] point P ; pointes A ; pointz B ; poyntes C ; *but grammar requires the singular.*

the *supplied from* CP ; AB *omit.*

§ 16, l. 5. AB *wrongly insert* the *before* Cancer ; CP *omit it.*

8. ilike] Ilyke G ; ilik P ; y-like C ; ilke AB ; see l. 7.

§ 17. *Latin rubric;* for *latitudinem* (as in M) read *longitudinem.* l. 17. heued B ; hed ACP ; see sect. 16, l. 3. The word "the" (rightly placed in BCMP) is, in A, wrongly placed before "aries" instead of before "ende."

23. the] þe C ; AB *omit.*

25. his] *miswritten* is *in* A ; *here, and in* ll. 18, 20, *and* 22.

§ 18, l. 2. on B ; upon C ; vpon MP ; vn A. Probably the form "vn" points to the reading "vpon" as being the correct one ; cf. note above to sect. 4, l. 6.

4. comeþ C ; comyþ P ; comth AB.

§ 19. *Latin Rubric;* for *orizon* (as in M) read *statio.*

7. comunely B; communely C; comuly A.

8. declineþ CP; declinet AB.

§ 20. *Latin Rubric;* the MS. (M) transposes the words *in* and *a,* having *a zodiaco in circulo,* which contradicts the sense.

2. his CP; is AB.

§ 22. *Latin Rubric;* for *centri* (as in M) read *contrade,* or *regionis.*

3. as hey is] as heiȝ is C; as hy is P; as hey as (*wrongly*) AB.

13. distance B] distaunce CP; destance A.

§ 23, l. 20. The figure "8" is omitted in AB. It is obviously re- ·quired.

22. than] A *omits;* thanne *inserted afterwards in* B.

§ 25. *Latin Rubric.* For *altitudinem,* M *has latitudinem,* an obvious error, due to the rubric of the preceding section.

3. the] *supplied from* B; AC *omit.*

15. CP *om.* And 10 minutes.

16. CP *om.* And minutes owt. *For* 51 degrees and 50 minutes, C *has* 52; þan is 52 degrees; *and* P *has* 52. þenne is .52. grees.

19. CP *om.* as y myght proue.

20. þe *supplied from* CP; AB *om.*

21. whaite] *so in* A; waite B; wayte C.

26. þe firste degre] 10 degrees C; 10 gree P.

27. 58 degrees and 10 Minutes] almost 56 C (*meaning* 56 degrees); almost .56. grees P.

28. almost 20] almost 18 C.

29. the] C *om.* and odde Minutes] CP *om.*

It thus appears that there is a second set of readings, involving a different calculation. The second set supposes the Sun to be in the 10th degree of Leo, his altitude to be 56°, and his declination 18°; the difference, viz. 38°, is the latitude. Either set of readings suits the. sense, but the one in the text agrees best with the former latitude, viz. 51°. 50'.

33. sonne C; *miswritten* sonnes AB.

36. *After* there, C *inserts* 38 grees, þat is; *and omits the words* of the pole, 51 degrees & 50 Minutes. But this is a mere repetition of the "height of the Equinoctial," and is obviously wrong. *After* pole, A *inserts* an that, *which is unmeaning, and omitted in* B.

39. nethere] neþerest CP.

§ 26, l. 8. The missing portion appears in MS. Bodley 619; I have not found it elsewhere. It is obviously correct, and agrees sufficiently closely with the conjectural addition by Mr Brae, in his·edition of Chaucer's Astrolabe, p. 48. He supplied the evident hiatus by the words—"A right circle or horizon have those people that dwell under the equinoctial line."

13. cenyth BC; cenytht A. A *inserts* the *between* 2 *and* wynteres; *absurdly.*

15. scheweþ CM; *miswritten* swewyth AB.

22. ouerkerueþ C; ou erkeruyht (*sic*) A; on ekir nyht (!) B; ouerkeruiþ P.

§ **27**, l. 2. the] *supplied from* BCPM; A *om.*

3. towchieth] towchiet A; towchet B; towchiþ P; towcheþ C; *see* l. 6.

§ **28**. *Latin Rubric.* The word *recto* is obviously wrong; read *obliquo,* and omit the last five words of the rubric.

2. thyn] *so in* B; þyn P; þin C; *miswritten* thyin A.

3. set] sett C; sete P; AB *omit.*

11. these] þese C; thise B; the A.

22. ende] heed A; heued C. In fact, *heed, heued,* or *hed* seems to be the reading of all the MSS. and printed copies, and may have been a slip of the pen in the first instance. The reading *ende* is, however, amply justified by its previous occurrence, four times over, in lines 9, 13, 16, 18. We thus have

Six Northern signs. From *head* of Aries to *end* of Virgo.

Six Southern signs. From *head* of Libra to *end* of Pisces.

Six Tortuous signs. From *head* of Capricorn to *end* of Gemini.

Six Direct signs. From *head* of Cancer to *end* of Sagittarius.

Opposite " sagittare " is written " sagittarie " in the margin of A, probably as a correction; but it is left uncorrected in l. 26.

§ **29**, l. 3. turne thanne] Turne þan C; turne the thanne AB; *where* " the " *is wholly superfluous; see* l. 8.

9. thow] þou C; two AB.

13. thorow] *so in* B; þorowȝ C; thoorw A. rewle] rule CP; *miswritten* rewles AB; *see* l. 9.

§ **30**. *English Rubric;* whether] wheþer CP; *miswritten* wherther AB.

6, 9. shewith] schewiþ P; scheweþ C; schewyþ M; swewith AB.

11. wey A; place C. *After* zodiak C *inserts*—for on þe morowe wol þe sonne be in a-noþer degre þan þan, & cetera; P *inserts*—For yn þe morowe wol þe sonne be yn an oþer gree, & norþer or souþer par aventure. Nothing can be plainer than that " the way of the sun " in this passage means the small circle formed by the sun's apparent path during a day; the text says expressly—" the wey wher as the sonne wente thilke day." We need not argue about the impossibility of a planet being found in " the way of the Sun " at midnight at the time of the Summer solstice, because Chaucer makes no assertion whatever here about the relative positions of the sun and planet; indeed, he carefully repeats " if " three times. He is only concerned with defining the phrase—" the latitude of a planet from the way of the sun "; and in every possible case, it is clear that a planet can be either (1) situate in the small circle called in the Latin rubric *cursus solis,* or (2) to the north of such a circle, or (3) to the south of such a circle. About this there need be no difficulty at all. It is all copied from Messahala.

§ **31**, l. 7. azymut] azymutz ABC; *but it is clear that the singular must be used,* as in sect. 32, l. 8. P *has* minute.

12. sowth B] þe souþe C; soth A.

§ **33**, l. 2. Azymut] Azymutz ABC; minutis P; *the same error as in* sect. 31, l. 7; *but see* sect. 32, l. 8. stondith] stondeþ C; shal stondith (*sic*) A; *where* shal *is over an erasure.*

3. in] yn P; ABC *omit.* It is of no consequence whether the word *in* be inserted or not; we find, on the one hand—"& in this same wyse maistow knowe by nyhte;" sect. 2, l. 6; and on the other—"the same wyse wyrke thow;" sect. 37, l. 14.

4. the nyht] *so in* AB; CP *om.* the; *and perhaps it is better omitted,* as in sect. 2, l. 6.

5. *After* north, B *inserts* or sowth; C *inserts* or southe.

6. is the sterre A; þe sterre stondeþ CP.

§ **34**. *English Rubric ;* latitude for] *so in* CP; latitude and for AB, *where* and *is superfluous, though perhaps it points to the reading* latitude *as* for.

5. is BC; his A.

6. towchith] touchiþ P; to which (*sic*) ABC; *see* sect. 27, l. 6. *In* A, *the word* assendente *is neatly written above* zodiak.

9. Astrolabie] Astrolobie A.

10. wheyther] wheþer CP; wheyth AB.

12. shewe] *so in* BP; schewe CM; swewe A.

13. this] þis P; thise AB; þese C; *moreover,* C *has* conclusiouns. But the singular seems intended; see l. 8.

§ **35**, l. 1. sterre BC; sterree A.

6, 7. whan C; wan AB (*twice*).

10, 13, 14. his C; is AB (*thrice*).

15. *After* west side, AB *add* & yf he be on the est syde, *a mere superfluous repetition ;* see l. 11.

17. sothly] soþly CP; *miswritten* he settes (!) AB.

18. hire Episicle] *so in* CP; *by an odd mistake,* AB *put* hire *after* manere, *instead of before* Episicle.

§ **36**, l. 3. remeue] Remewe CP; remue AB.

5, 7, 10, 16. I leave the spelling *howys* (or *howis*) as it stands in the MS.; see *house* in l. 12; *hows* in l. 13; *hous* (as in C) in l. 15.

16. *Here* A *inserts* the *before* nadir; it might have omitted, as in ll. 13, 14, 15, and 17. Indeed, MS. B omits it.

§ **37**, l. 6. the degree] þe degree C; thee degree A. assendeth] ascendeþ C; assendet A; assendent B.

9. than] þan C; AB *omit.*

10. is] AB *omit; but it is obviously wanted;* C *varies here.*

12. 12 howse next] 12 hous next C; howses nex (*sic*) AB.

13. thanne] þan C; fro (!) B; A *omits.* howse] hous C; howses AB.

16. AB *absurdly insert* fro *before* the bygynnyng.

17. the] þe C; AB *omit.*

§ **38**, l. 1. warpyng MP; werpynge C; weripinge (*sic*) A; wernipinge (*sic*) B.

2. a CP; AB *omit.*

3, 4. an euene C; a euene AB (*twice*).

7. fro the centre; i. e. *above* the centre. The length of the pin, measured from the centre in which it is inserted, is to be not more than a quarter of the diameter, or half the radius. This would make the ratio of the gnomon to the shadow (or radius) to be one-half, corresponding to an altitude a, where tan $a = \frac{1}{2}$; i. e. to an altitude of about $26\frac{1}{2}°$. As Chaucer talks about the sun's altitude being $25\frac{1}{2}°$ at about 9 o'clock, at the time of the equinoxes (sect. 3), there is nothing that is particularly absurd in the text of this section. For Mr Brae's conjectural emendations, see p. 56 of his edition.

15. tak thanne] *so in* P; tak me thanne AB; take me þan C. But there seems no sufficient reason for thus inserting *me* here. Cf. "Tak a rond plate," l. 1; "tak than a compas," l. 14; "tak ther thy lyne," l. 17.

§ 39. At this point MS. A, which has so far, in spite of occasional errors of the scribe, afforded a very fair text, begins to break down; probably because the corrector's hand has not touched the two concluding sections, although section 40 is much less corrupt. The result is worth recording, as it shews what we may expect to find, even in good MSS. of the Astrolabe. The section commences thus (the obvious misreadings being printed in italics) :—

"This lyne Meridional ys but a Maner descripcion *or the* ymagined, that passeth vpon the pooles of þis *the* world And by the cenyth of owre heued / And hit is *the same* lyne Meridional / for in what place þat any maner man [*omission*] any tyme of the yer / whan that the sonne *schyneth ony thing* of the firmament cometh to his verrey *Middel lyne of the* place / than is hit verrey Midday, þat we clepen owre noon," &c.

It seems clear that this apparent trash was produced by a careless scribe, who had a good copy before him; it is therefore not necessary to reject it all as unworthy of consideration, but it is very necessary to correct it by collation with other copies. And this is what I have done.

MS. B has almost exactly the same words; but the section is considerably better, in general sense, in MSS. C and P, for which reason I here quote from the former the whole section.

[Rawl. MS. Misc. 1370, *fol.* 40 *b.]*

Descripcioun of þe meridional lyne, of þe longitudes and latitudes of Citees and townes, as wel as of a (*sic*) clymatz.

39. *conclusio.* This lyne meridional is but a maner discripcion or lyne ymagyned, þat passeþ vpon þe pooles of þis worlde, and by þe Cenith of oure heued. ¶ And yt is cleped þe lyne meridional, for in what place þat any man ys at any time of þe ȝere, whan þat þe sonne by meuynge of þe firmament come to his uerrey meridian place / þan is it þe uerrey mydday þat we clepe none, as to þilke man. And þerefore is yt cleped þe lyne of mydday. And no*ta*, þat euermo of any .2.

citees or of 2 townes, of which þat oo towne a-procheþ neer þe est þan
doþ þe oþer towne, trust wel þat þilke townes han diuerse meridians.
Nota also, þat þe arche of þe equiuoxial, þat is contened or bownded
by-twixe þe two meridians, is cleped þe longitude of þe towne. ¶ &
ȝif so be / þat two townes haue I-like meridian or one merydian,
¶ Than ys þe distaunce of hem boþe I-like fer from þe est, & þe contra-
rye. ¶ And in þis maner þei chaunge not her meridyan, but soþly, þei
chaungen her almykanteras, For þe enhaunsynge of þe pool / and þe
distaunce of þe sonne. ¶ The longitude of a clymate ys a lyne
ymagyned fro þe est to þe west, I-like distaunte fro þe equinoxial.
¶ The latitude of a clymat may be cleped þe space of þe erþe fro þe
by-gynnynge of þe first clymat unto þe ende of þe same clymat / euene-
directe a-ȝens þe pool artyke. ¶ Thus seyn somme auctours / and
somme clerkes seyn / þat ȝif men clepen þe latitude of a contrey,[1] þe
arche mer[i]dian þat is contened or intercept by-twixe þe Cenyth & þe
equinoxial ; þan sey þei þat þe distaunce fro þe equinoxial unto þe ende
of a clymat, [2]euene a-gaynes þe pool artik, is þe latitude off þat climat[2]
forsoþe.

The corrections made in this section are here fully described.

1. of lyne P ; of a line I ; or lyne C ; or the AB.

2. þis] þis the AB, *absurdly ;* CP *omit* the, *rightly.*

3. ycleped the] y-clupid þe P ; cleped þe C ; the same (*sic*) AB.

4. ys at ; *supplied from* PCI ; AB *omit.*

5. by moeuyng] by meuynge C ; by mevyng PI ; schyneth ony
thing (*sic*) A ; schyned eny thing B ; *for the spelling* moeuyng, *see sect.*
35, l. 5.

6. meridian CP ; meridianale I ; Middel lyne of the (*sic*) AB.

8. 2 citees CI ; too citees P ; any lynes (*sic*) AB.

9. aprochith] a-procheþ C ; aprochiþ P ; *miswritten* aprochid AB.
more toward] neer C ; ner P ; neerer I ; thoward AB.

11. conteyned I ; conteynyd P ; contened C ; consideered (*sic*) A ;
contined B.

13. yf P ; ȝif C ; if it I ; AB *omit.* N.B. It would have been
better to have used the spelling *yif,* as the word is commonly so spelt
in A.

21. same CPI ; seconde AB. The reading *same* is right ; for the
" latitude of a climate " means the breadth of a zone of the earth, and
the latitude of the first climate (here chosen by way of example) is the
breadth as measured along a line drawn perpendicular to the equator,
from the beginning of the said first climate to the end of *the same.*
The words " euene-directe agayns þe poole Artik " mean in the direction
of the North pole ; i. e. the latitude of a climate is reckoned from its
beginning, or *southernmost* boundary-line towards the end of the same,
viz. its *northern* boundary-line.

[1] Here insert—[þey mene]—which CP omit.

[2] The words from *euene* to *climat* are added at the bottom of the page in
the MS.

22. þe poolo Artik P; þe pool ártyke C; the polc artike I; from north to south AB. Observe that this singular error in A, " euene directe agayns from north to south," probably arose from a confusion of the text " euene directe agayns þe poole Artik " with a gloss upon it, which was " from north to south." It is important as throwing light on the meaning of the phrase, and proving that the interpretation of it given above (note to l. 21) is correct.

24. intercept CP; intercepte I; except (*over an erasure*) AB.

The only reading about which there is any doubt is that in line 18, which may be either " illike distant by-twene them alle " (A), or " I-like distaunte fro þe equinoxial " (C). But it is immaterial which reading be adopted, since *Illike-distant* is here used merely in the sense of *parallel*, and the boundaries of the climates are parallel both to one another, and to the equinoctial. The climates themselves were of different breadths.

§ **40**, l. 4. this samples AB; þese ensamples C. *For* this *read* thise *or* these.

5. for sothe] *miswritten for* sonne AB; in general C; yn special P; *the reading* sonne *points to* sothe, *and makes it very probable that* for sothe *is the true reading.*

6. longitude] þe longitude C; latitude AB (*absurdly*); see l. 11. Perhaps we should read " the longitude " ; but it is not very material.

7. planete; *miswritten* that A, *but corrected to* planete *in the margin;* C *has* planete, *correctly.* The figure 6 is omitted in C; so are all the other figures further on. him] hir C.

8. I tok] Than toke I C. 8, 15. 2 degrees A; 3 degrees B.

10. Than tok I] Than toke I C; *for* tok AB *wrongly have* stykke, *afterwards altered to* stokke *in* A. the] *supplied from* C, *which has* þe ; AB *omit.*

15. hauy A; haue I C.

22. the] þe C; AB *omit.*

25. prikke] prickes C; *perhaps* prikkes *would be a better reading.*

27. AB *omit the figure* 2 ; *but see* l. 8.

30. in alle] in al C; A *has* septentrionalle, *an obvious mistake for* septentrional in alle, *by confusion of the syllable* " al " *in the former with* " al " *in the latter word;* B *has* septentrional, *omitting* in alle.

33. sothly] *so in* B; soþly C; *miswritten* sothtly A ; see l. 30. signes C] tymes AB (*wrongly*) ; see l. 30.

39. hauy AB ; haue I C.

43. *Perhaps* euene *before* of *should be omitted, as in* C. AB *have* in the ende euene ouer of thee, *where* euene ouer *is repeated from the former part of the line.*

44. F endlang] F endlonge C ; A euene AB ; *but see* l. 22.

45. A *omits* of *and* degrees, *yet both are required;* BC *omit* of 3 degrees *altogether.*

47. til] tyl þat C ; tho AB (*absurdly*). saw] sey C ; may AB ; *see* l. 27.

53. hire] his ABC. a] ABC *omit.*

54. *At the word* howrc *four of the best* MSS. *break off,* viz. MSS. ABCE, *although* E *adds one more section, viz.* sect. 46; *others come to a sudden end even sooner, viz.* MSS. DFGHK. *But* MS. P *carries us on to the end of* sect. 43, *and supplies the words*—þu shalt do wel ynow.

§ **41**, 6. betwen] be M (*wrongly*); by-twyx L; *see* l. 5.

7. M *inserts* & *before* to þe altitude; *a mere slip.* For; *miswritten* Fro M. þridde; *miswritten* ridde M.

11. poynt L; *miswritten* poyn M.

12. LM *wrongly place of* after *þe* hey3t *instead of before it;* see the footnote.

§ **42**, l. 2. see] *so in* L; *miswritten* sette M; *see* sect. 41, l. 4.

3. y] I L; M *omits.*

7. M *omits* as, *above, and* is þe; L *has* 12 passethe 6 the.

10. seest] *so in* L; *miswritten* settest M; *cf.* l. 2; P *has* sixt (*a common old form for* seest).

11. 60] LN; sexe M.

12. M *omits from* 10 is *to* 10 fete, *which is supplied from* NLP.

13. For] *so in* LN; fro M.

14. *For* 2, M *has* 6.

15. *For* 3, M *has* 4; *for* 2, M *has* 6; *for* 6, M *has* 2; *and the words* and 3 is 4 partyes of 12 *are omitted, though* L *has*—& 4 is the thrid partye of 12.

17. betwen] by-twene L; bitwixe P; *miswritten* be M; *cf.* sect. 41, 6.

18. þre] 3 LP; *miswritten* þe M.

§ **43**. Rubric, *Vmbra Versa;* obviously a mistake for *Recta.* The error is repeated in l. 1. LP rightly read *Recta.*

3. M *omits* 1, *which is supplied from* LP; *see* l. 5.

4. difference] diffrense M; *cf.* sect. 42, 17.

10. *After* heythe, LN *add* to thyn eye. *In place of lines* 8—10, P *has*—& so of alle oþer, &c.

§ **44**. From MS. Digby 72 (N). Also in LMO.

2. fro] *so in* LO; for M.

3. in-to] *so in* L; in M. for] *so in* O; fro M.

6. 3eris M; LNO *omit.*

7. tabelis NO; table M; tables L.

8. where L; qwere O; wheþer N.

9. loke LM; N *omits.*

10. NM *omit from* or what *to* or; *supplied from* O, *which has*—or qwat nombre þat euere it be, tyl þe tyme þat þou come to 20, or 40, or 60. *I have merely turned* qwat *into* what, *as in* L, *which also has this insertion.*

12. wreten N; *the alteration to* wryte *is my own;* see l. 22.

vnder] *so in* L; vndirneþe M.

13. to-geder] too-geder M; *miswritten* to 2 degreis N; to the 2 degrees L.

14. hast M; *miswritten* laste N; last L.

15. þat; *supplied from* M; LN *omit.*

16. *For* 1 (*as in* M) LN *have* 10.

20. to-gedere M; to the degreis N; 2 grees O; to degrees L.

21. þat; *supplied from* M; LNO *omit.*

lasse] passid LNO; M *omits.* Of course *passid* is wrong, and equally of course *lasse* is right; see ll. 5, 6 above, and l. 24 below.

24. þat] *so in* L; þat MO; if hit N.

25. entrynge] entre M; entre L.

26. þer] *so in* M; *miswritten* the ȝere N; the ȝeer L.

28. merydie LM; merdie N.

30. for LM; fro N (*twice*).

32. thaȝthe N; haue tauȝt M; haue tawȝt O; haue tauht L.

33. þe; *supplied from* M; LNO *omit.*

34. *with* þe] *so in* M; wyche N; *see* l. 36.

38. in-to N; yn M.

§ **45.** From MS. Digby 72 (N); also in LO; but not in M.

4. that] the L; þe O.

6. I-wryton] wrytoun O; Iwyton N. *But* L *has* I wold wyttyn, *and I would therefore, on second thoughts, propose to read*—I wolde wyten precise my rote.

7. 1397] *miswritten* 1391 LN; O *has* 1391, *corrected to* 1397; see l. 3.

10. soȝth N; sowte O; sowthe L.

12. vnder N; vndyr-nethe O; vndre-nethe L.

18. oþer in any oþer tyme or monyth N; or any oder tymys or monthys O; or in eny other moneth L.

24. adde] *supplied from* L; NO *omit.* There is no doubt about it, for see l. 14.

27. wete þe] *so in* O; wete thi L; *miswritten* with thy N; see l. 17.

31. and] *supplied from* LO; N *omits.*

32. abowe N; aboue LO.

36. lewyth N; leuyth LO.

§ **41a.** This and the remaining sections are almost certainly spurious. The last occurs in MS. Bodley 619 (E) only; the others are in LMN, the first (41a) being also found in O. The text of 41a—42b is from M.

3. hast] *supplied from* L; M *omits.*

§ **42a,** 1. heyth by þy N; heyth by the L; M *om.*

4. lyk] lykk M; L *omits.* mete] mette M; mett L.

9. ys] is L; *miswritten* hys M.

§ **43a,** 1. nat] nott L; M *omits;* see the footnote. In the rubric, M has *versam;* but L has the rubric—*Vmbra Recta.*

§ **42b,** 5. as] *so in* L; *miswritten* & M.

6. 4 *is supplied from* L; M *omits.*

§ **46,** 9. it] E *omits.*

14. maist] E *omits.*

ADDITIONAL NOTES.

PROLOGUE. l. 7. *suffisaunt*, sufficiently good. In the best instru-ments, the Almicanteras, or circles of altitude, were drawn at distances of one degree only; in less carefully-made instruments, they were drawn at distances of two degrees. The one given to his son by Chaucer was one of the latter; see Part II. sect. 5.

10. *a certein*, i. e. a certain number; but the word *nombre* need not be repeated; cf. *a certein holes*, Pt I. sect. 13, l. 2, and see the very expression in the Milleres Tale, l. 7.

20. *suffise*, let them suffice.

58. "Nicolaus de Lynna, i. e. of Lynn, in Norfolk, was a noted astrologer in the reign of Edward III., and was himself a writer of a treatise on the Astrolabe. See Bale—who mentions 'Joannes Sombe' as the collaborateur of Nicolaus—'Istos ob eruditionem multiplicem, non vulgaribus in suo Astrolabio celebrat laudibus Galfridus Chaucer poeta lepidissimus;' BALE (edit. 1548), p. 152."—Note by Mr Brae, p. 21 of his edition of the Astrolabe.

Warton says that "John Some and Nicholas Lynne" were both Carmelite friars, and wrote calendars constructed for the meridian of Oxford. He adds that Nicholas Lynne is said to have made several voyages to the most northerly parts of the world, charts of-which he presented to Edward III. These charts are, however, lost. See Hak-luyt's Voyages, i. 121, ed. 1598; Warton, Hist. E. P. ii. 357; ed. 1871.

Tyrwhitt, in his Glossary to Chaucer, s. v. *Somer*, has the following. "The Kalendar of John Somer is extant in MS. Cotton, Vesp. E. vii. It is calculated for 140 years from 1367, the year of the birth of Richard II., and is said, in the introduction, to have been published in 1380, at the instance of Joan, mother to the king. The Kalendar of Nicholas Lenne, or Lynne, was calculated for 76 years from 1387. Tanner in v. *Nicolaus Linensis*. The story there quoted from Hakluit of a voyage made by this Nicholas in 1360 *ad insulas septentrionales antehac Europæis incognitas*, and of a book written by him to describe these countries *a gradu .54. usque ad polum*, is a mere fable: as appears from the very authorities which Hakluit has produced in support of it." It seems probable, therefore, that the "charts" which Warton says are "lost" were never in existence at all. The false spelling "Some" no doubt arose from neglecting the curl of contraction in Som*ere*.

PART I. § **5**, l. 5. *the remanant,* &c. i. e. the rest of this line (drawn, as I said,) from the foresaid cross to the border. This appears awkward, and we should have expected " fro the forseide *centre*," as Mr Brae suggests ; but there is no authority for making the alteration. As the reading stands, we must put no comma after " this lyne," but read on without a pause.

7. *principals.* It is not unusual to find adjectives of French origin retaining *s* in the plural; only they commonly *follow* their nouns when thus spelt. Cf. *lettres capitals,* i. 16. 8. On the other hand, we find *principal cerkles,* i. 17. 33.

§ **7**. 4. *nowmbres of augrym ;* Arabic numerals. The degrees of the border are said to contain 4 minutes *of time,* whilst the degrees of the signs are divided into minutes and seconds of angular measurement, the degrees in each case being the same. There is no confusion in practice between these, because the former are used in measuring time, the latter in measuring angles.

§ **8**. 9. *Alcabucius ;* i. e. (says Warton, Hist. E. P. ii. 357, ed. 1871) Abdilazi Alchabitius, whose *Introductiorium ad scientiam judicialem astronomiæ* was printed in 1473, and afterwards. Mr Brae quotes the very passage to which Chaucer refers, viz. " Et unumquidque istorum signorum dividitur in 30 partes equales, quæ gradus vocantur. Et gradus dividitur in 60 minuta ; et minutum in 60 secunda; et secunda in 60 tertia ; similiterque sequuntur quarta ; similiter et quinta ; ascendendo usque ad infinita ; " Alchabitii Differentia Prima.

These minute subdivisions were never used ; it was a mere affectation of accuracy, the like of which was never attained.

§ **10**. 5. *in Arabyens,* amongst the Arabians. But he goes on to speak only of the Roman names of the months. Yet I may observe that in MS. Ii. 3. 3, at fol. 97, the Arabian, Syrian, and Egyptian names of the months are given, as well as the Roman.

§ **16**. 12. *& euery Minut* 60 *secoundes ;* i. e. every minute contains 60 seconds. The sentence, in fact, merely comes to this. " Every degree of the border contains four minutes (*of time*), and every minute (of time) contains sixty seconds (of time)." This is consistent and intelligible. Mr Brae proposes to read "*four* seconds " ; this would mean that " every degree of the border contains four minutes (of time), and every minute (*of the border*) contains four seconds (of time)." Both statements are true ; but, in the latter case, Chaucer should have repeated the words " of the bordure." However this may be, the proposed emendation lacks authority, although the reprint of Speght changed " lx " into " fourtie," which comes near to " four." But the reprint of Speght is of no value at all. See Mr Brae's preface, p. 4, for the defence of his proposed emendation.

§ **17**. 6. *Ptholome.* The John's MS. has *ptolomeys almagest.* "*Almagest,* a name given by the Arabs to the μεγάλη σύνταξις, or *great collection,* the celebrated work of Ptolemy, the astronomer of Alexandria [floruit A.D. 140—160]. It was translated into Arabic about the year

A.D. 827, under the patronage of the Caliph Al Mamun, by the Jew Alhazen ben Joseph, and the Christian Sergius. The word is the Arabic article *al* prefixed to the Greek *megistus*, 'greatest,' a name probably derived from the title of the work itself, or, as we may judge from the superlative adjective, partly from the estimation in which it was held."—English Cyclopædia; Arts and Sciences, i. 223. The Almagest "was in thirteen books. Ptolemy wrote also four books of judicial astrology. He was an Egyptian astrologer, and flourished under Marcus Antoninus. He is mentioned in the Sompnour's Tale, l. 1025, and the Wif of Bathes Prologue, l. 324."—Warton, Hist. E. P. ii. 356, ed. 1871. The word *almagest* occurs in the Milleres Tale, near the beginning, and twice in the Wif of Bathes Prologue.

Chaucer says the obliquity of the ecliptic, according to Ptolemy, was 23°. 50′. The *exact* value, according to Ptolemy, was 23°. 51′. 20″; *Almagest*, lib. i. c. 13. But Chaucer did not care about the odd degree, and gives it nearly enough. See note to ii. 25. 18.

8. *tropos*, a turning; Chaucer gives it the sense of *agaynward*, i. e. in a returning direction.

14. The equinoctial was supposed to revolve, because it was the "girdle" of the *primum mobile*, and turned with it. See note below to l. 27.

14. "As I have shewed thee in the solid sphere." This is interesting, as shewing that Chaucer had already given his son some lessons on the motions of the heavenly bodies, before writing this treatise.

26. *angulus*. We should rather have expected the word *spera* or *sphera;* cf. "the sper solide" above, l. 15.

27. "And observe, that this first moving (*primus motus*) is so called from the first movable (*primum mobile*) of the eighth sphere, which moving or motion is from East to West," &c. There is an *apparent* confusion in this, because the *primum mobile* was the *ninth* sphere; but it may be called the movable of the eighth, as *giving motion to it*. An attempt was made to explain the movements of the heavenly bodies by imagining the earth to be in the centre, surrounded by a series of concentric spheres, or rather shells, like the coats of an onion. Of these the seven innermost, all revolving with different velocities, each carried with it a planet. Beyond these was an eighth sphere, which was at first supposed to be divided into two parts, the inner part being the *firmamentum*, and the outer part the *primum mobile;* hence the *primum mobile* might have been called "the first moving of the eighth sphere," as accounting for the more important part of the motion of the said sphere. It is simpler, however, to make these distinct, in which case the eighth sphere is the *firmamentum* or *sphæra stellarum fixarum*, which was supposed to have a very slow motion from West to East round the poles of the *zodiac* to account for the precession of the equinoxes, whilst the ninth sphere, or *primum mobile*, whirled round from East to West once in 24 hours, carrying all the inner spheres with it, by which means the ancients accounted for the diurnal revolution.

This ninth sphere had for its poles the north and south poles of the
heavens, and its "girdle" (or great circle equidistant from the poles)
was the equator itself.   Hence the equator is here called the "girdle of
the first moving."   As the inner spheres revolved *in an opposite direction*,
to account for the *forward* motion of the sun and planets in the ecliptic
or near it, the *primum mobile* was considered to revolve in a *backward*
or *unnatural* direction, and hence Chaucer's apostrophe to it (Man of
Lawes Tale, 295)--

> " O firste moeuyng cruel firmament,
>   With thy diurnal sweigh that crowdest ay
>   And hurlest al from Est til Occident,
>   That naturelly wolde holde another way."

That is—" O thou *primum mobile*, thou cruel firmament, that with thy
diurnal revolution (or revolution once in 24 hours round the axis of the
equator) continually forcest along and whirlest all the celestial bodies
from East to West, which *naturally* would wish to follow the course of
the sun in the zodiac from West to East."   This is well illustrated by
a sidenote in the Ellesmere MS. to the passage in question, to this
effect :—"Vnde Ptholomeus, libro i. cap. 8.   Primi motus celi duo
sunt, quorum vnus est qui mouet totum semper ab Oriente in Occidentem
vno modo super orbes, &c.   Item aliter vero motus est qui mouet orbem
stellarum currencium contra motum primum, videlicet, ab Occidente in
Orientum super alios duos polos." [1]   That is, the two chief motions are
that of the *primum mobile*, which carries everything round from East to
West, and that of the fixed stars, which is a slow motion from West to
East round the axis of the zodiac, to account for precession.   This
exactly explains the well-known passage in the Frankeleines Tale (C.
T. ed. Tyrwhitt, 11592)—

> " And by his eighte speres in his werking,
>   He knew ful wel how fer Alnath was shove
>   Fro the hed of thilke fix Aries above
>   That in the ninthe spere considered is."

Here the eight spheres are the eight inner spheres which revolve round
the axis of the zodiac in an easterly direction, whilst the ninth sphere,
or *primum mobile*, contained both the theoretical or *fixed* first point
of Aries from which measurements were made, and also the *signs* of
the zodiac as distinct from the *constellations*.   But Alnath, being an
actual star, viz. a Arietis,[2] was in the *eighth* sphere ; and the distance
between its position and that of the first point of Aries at any time
afforded a measure of the amount of precession.   Mr Brae rightly re-

---

[1] This is doubtless quoted from some gloss upon Ptolemy, not from the
work itself.   The reference is right, for the "motus celi" are discussed in the
Almagest, lib. i. c. 8.
[2] This star (a Arietis) was on the supposed horn of the Ram, and hence its
name ; since *El-nâtih* signifies " the butter," and "El-nath " is " butting " or
"pushing."   See Ideler, Die Bedeutung der Sternnamen, p. 135.

marks that Tyrwhitt's readings in this passage are correct, and those of Mr Wright and Mr Morris (from the Harleian MS.) are incorrect.

It may be as well to add that a later refinement was to insert a crystalline sphere, to account for the precession ; so that the order stood thus : seven spheres of planets ; the eighth, of fixed stars ; the ninth, or crystalline ; the tenth, or *primum mobile ;* and, beyond these, an empyræan or theological heaven, so to speak, due to no astronomical wants, but used to express the place of residence of celestial beings.[1]   Hence the passage in Milton, iii. 481.

> "They pass the planets seven, and pass the fix'd,
> And that crystalline sphere whose balance weighs
> The trepidation talk'd, and that first mov'd."

i. e. They pass the seven planetary spheres ; then the sphere of fixed stars ; then the crystalline or transparent one, whose swaying motion or libration measures the amount of the precession and nutation so often talked of ; and then, the sphere of the *primum mobile* itself.   But Milton clearly himself believed in the Copernican system ; see Paradise Lost, viii. 121—140, where the *primum mobile* is described in the lines—

> "that swift
> Nocturnal and diurnal rhomb supposed,
> Invisible else above all stars, the wheel
> Of day and night."

§ **18.** 8. *compowned by* 2 *&* 2.   This means that in the *best* astrolabes, *every* almicantarath for every degree of latitude was marked ; as may be seen in Metius.   In others, including the one given by Chaucer to his son, they were marked only for every other degree.   See Part II. sect. 5, l. 2.

§ **19.** 7. *cenyth*, as here used, has a totally different meaning from that of *senyth*, in l. 1 above.   The *senyth* in l. 1 is what we still call the *senyth ;* but the *cenyth* in l. 7 means the point of the horizon denoting the sun's place in azimuth.   Contrary to what one might expect, the *latter* is the true original meaning, as the word *zenith* is corrupted from the root of the word which we now spell *azimuth.*   The Arabic *as-samt* is a way or path ; *al-samt,* a point of the horizon, and, secondly, an azimuthal circle.   The plural of *al-samt* is *assumût,* whence *azimuth.* But *zenith* is a corruption of *semt,* from *samt al-ras,* the Arabic name of the vertex of heaven (*râs* meaning *a head*) ; and the qualifying *al-ras,* the most important part of the phrase, has been improperly dropped.   So far from the reading *cenyth* being wrong here, it is most entirely right, and may be found in the same sense in Messahala.   See p. 41, footnote. For *cenyth,* some late copies have *signet,* evidently taken from the

---

[1] Well expressed by Dante, *Parad.* xxx. 38—

> "Noi semo usciti fuore
> Del maggior *corpo* al ciel ch'é pura luce."

Dante, like Chaucer, makes the eighth sphere that of fixed stars, and the ninth the *primum mobile* or swiftest heaven (*ciel velocissimo*); *Parad.* xxvii. 99.

Latin word *signum*. They make the same mistake even in l. 10 of section 18.

§ **21.** 4. *sterres fixes*, fixed stars; here the *s* again appears in a plural adjective of French derivation. In MSS. Ii. 3. 3 and Ii. 1. 13 in the Cambridge University Library, is an interesting list of the 49 stars most usually placed upon the Astrolabe, which I have printed in the Preface to this volume. The stars which are represented by the points of the tongues in Fig. 2 are the same as those in the diagram from which Fig. 2 is copied, the original of which is in MS. A. I have slightly altered the positions of the points of the tongues, to make them somewhat more correct. The following is the list of the stars there shewn ; most of their names are written in the MS. Cf. footnote on p. 12.

Within the Zodiac. In *Aries*, Mirach, or β Andromedæ, shown by a short tongue above Aries ; in *Taurus*, Algol, or β Persei, as marked ; in *Libra*, Aliot or Alioth, i. e. ε Ursæ Majoris (the third horse, next the cart, in Charles's Wain), as marked ; also Alramech, Arcturus, or α Boötis, shewn by the tongue projecting above Libra ; in *Scorpio*, Alpheta, Alphecca, or α Coronæ Borealis, as marked ; in *Sagittarius*, Raz Alhagus, or α Ophiuchi, near Alpheta ; in *Capricornus*, Altair or α Aquilæ and Vega or α Lyræ, as marked, whilst near Vega is the unmarked Arided, or α Cygni ; and in *Pisces*, Markab or α Pegasi.

Without the Zodiac. In *Aries*, under *Oriens*, the slight projection marks β Ceti or Deneb Kaitos, the Whale's Tail, and the next curiously shaped projection (with side-tongues probably referring to other stars) means Batukaitos or Batnkaitos, the Whale's Belly, apparently ζ Ceti ; next come the long tongue for Menkar or α Ceti, the Whale's Nose ; the star Aldebaran or Bull's Eye, α Tauri ; Rigel or β Orionis, Orion's Foot ; Alhabor or Sirius, the Dog-star, marked by a rude drawing of a dog's head, the star itself being at the tip of his tongue; then Algomeisa, Procyon, or α Canis Minoris, marked by a tongue pointing to the left, whilst the long broad tongue pointing upwards is Regulus, Kalbalased, or α Leonis ; the small tongue above the letter I in the border is Alphard or Cor Hydræ. Above *Occidens*, in *Libra*, the first tongue is Algorab or δ Corvi, and the next Spica Virginis or Azimech ; close to the 8th degree of Scorpio is α Libræ, and close to the beginning of Sagittarius is a small head, denoting the Scorpion, at the tip of the tongue of which is the bright Kalbalacrab or Antares. The last, a projection below the letter X, is Deneb Algebi or the Goat's Tail, i. e. δ Capricorni.

7. That is, the little point at the end of each tongue of metal is technically called the " centre " of the star, and denotes its exact position.

9. The stars of the North are those to the North of the *zodiac*, not of the *equator*.

12. *Aldeberan*, &c. ; the stars Aldebaran (α Tauri) and Algomeisa (α Canis Minoris) are called stars of the south, because they are to the south of the ecliptic ; but as they are meanwhile (see Fig. 2) also to the north of the equator, they of course rise to the N. of the Eastern point of the horizon. The longitude of stars was always measured

along the ecliptic, which is denoted in Fig. 2 by the outermost circle of the metal ring on which the names of the signs are written.

In one of the tracts in MS. G (dated A.D. 1486), p. 30, we find "Aldebaran, in the first gre of geminis (sic), of the nature of Mars and Venus;" and "Algomeisa, canis minor, in the xvij gre of Cancer, of the nature of Mars and Mercury."

28. *Amiddes*, &c. Observe that the Ecliptic line, though *in the midst* of the *celestial* zodiac, a belt 12° broad, is on the *outer edge* of the zodiac as shewn in the astrolabe, which is only 6° broad and shews only the northern half of that belt. The "way of the sun" is elsewhere used of the sun's apparent *diurnal* path (see Part ii. sect. 30); but it here probably refers, as is more usual, to the *annual* path.

33. *streitnes*, narrowness, closeness, smallness of size. In Fig. 2, I have marked every degree in the southern half of the zodiac, but only every *fifth* degree in the northern, in order to avoid an appearance of crowding in so small a figure. In Chaucer's own astrolabe, every *other* degree was marked all round.

38. Here Chaucer gives at least three reasons for the name of "zodiac." The true one is the second, "for that the sterres that ben there fixed ben disposed in signes of bestes, or shape like bestes." But these imaginary shapes are very absurd and arbitrary.

48. Not only the influences here assigned to the signs, but others due to planets, may be found in "Porphyrii Philosophi introductio in Claudii Ptolomæi opus de affectibus astrorum," fol. Basileæ, n. d. p. 198. I here add a few extracts from the MS. in Trinity College, Cambridge (marked R. 15. 18), to shew the nature of the old astrology. I choose them with especial reference to Aries. The other signs are spoken of in a similar manner. "It is principally to be considered that the signes of hevyn haue theire strenght and propre significacioun vpon the membris of eny man; as, Aries hath respect to the hed, taurus to the neck, geminis (sic) the Armys, Cancer the brest, leo the hert, virgo the bowels, &c; as it shall shew in the Chapiters folowyng.[1] Secundarily it is to be noted that plotholomee (sic) saith, that to touche with instrument of yroun while the mone is in the signe of the same membre, is for to be dred; let the surgen beware, and the letter of blode, let hym be aferd to touche that membre *with* yrene, in the which the mone shal be."—MS. G (see the preface); Tract C. p. 12.

"Thenne Aries hath respect to the hed; And this signe is hote and dry, fiery & colerik. Saturne hath ij witnes in Ariete, a triplicitate and a terme. Jubiter also hath ij, a triplicitate and a terme. Mars hath iij testimonials or iij fortitudis in Ariete, A hows, A face, and A terme. The sonne hath iij fortitudis in Ariete, *scilicet*, an exaltacioun, a triplicite, and a face. Venus hath ij testimonials, A terme and a face. Mercury hath one testymony, that is to sey, a terme. And luna

[1] From this same MS. I have copied the figure of a man shewn in Fig. 19. The copy was hurriedly made, and is by no means a good one; but it may serve to shew the manner of denoting the influence of the signs upon a man's body.

in Ariete hath no testimoniall.   For the which it is to know, that the influens of the planet*is* may be fortyfied v maner of wayes.   And these v maner be called v fortitud*is* of planetis, or testimonials, which be these : *domus, exaltacio, triplicitas, terminus,* and *facies. Domus* gevith to a planet v fortitud*is*; And a planet in his hows is lyke a kynge in his hall, And in the high trone of his glorie.   A planet in his *exaltacioun* is lyke a kynge when he is crowned.   A planet in his *triplicite* is like a kynge in hono*ur*, Amonge his sencible people.   A planet in his *terme* is As a man*n* amonge*s* his kynnesmen*n* And fryndis.   *Facies* gyvith to A planet that thyng the which rowme gyvith to a maistre.   Wherfore *facies* gyvith only on fortitude, *Terminus* ij, *Triplicitas* iij, *Exaltacio* iiij, And *domus* v.   And for the more clere declaracio*n*, the dignytes of planett*is* in signes be comprehendid in this figure ensuynge, &c.[1] "— Same MS., Tract C. p. 13.

" The dygnytes of planet*is* in the signes, most speciall they be to be noted in iudicials.   When the mone is *in Ariete*, it is not gode, but vtterly to be exshewed, both for seke And disesid, for to shafe theire hede or to boist in the eris or in the nek ; nor loke þou let no blode in the vayn of the hede.   How-be-it, benyficiall it is to begynne euer*y* worke that þou woldest bryng aboute son*e*.   But that thynge that is stabill ought to be eschewed.   In this signe it is necessary to dele with noble estat*is* And rich men, And for to go in-to A bayne [*bath*].[2]"—Same MS., Tract C. p. 14.

52.   See Prologue, l. 68.   As the zodiak is here called a part of the eighth sphere, so we have been before told that the equinoctial is the girdle of the ninth sphere ; see note above to sect. 17. l. 27.

54.   *euene parties,* equal parts.   That is, the equinoctial bisects the zodiac.   But the northern half *looks* much smaller than the southern on the Astrolabe, owing to the manner in which the zodiac is there represented, viz. by projection on the plane of the equator.

Part II. § 1.   *Rubric.   hir cours.*   The gender of the sun was feminine in Anglo-Saxon, and that of the moon masculine ; but in Chaucer's time, the gender was very variable, owing to the influence of Latin and French.

§ 3.   Between sections 2 and 3, a section is inserted in the late

---

[1] Here follows a table, shewing that, in *Aries,* the value of *Saturn* is 5, of *Jupiter* 5, &c. ; with the values of the planets in all the other signs. The value 5, of Saturn, is obtained by adding a *triplicite* (value 3) to a *terme* (value 2), these being the " witnesses." of Saturne in Aries ; and so on throughout.

[2] So on p. 12 of another tract (D) in the same MS., we find—

Aries calid*u*m & suc*u*m ; bon*u*m.

Nill capiti noceas, Aries cum luna refulget,
De vena minuas & balnea tuti*us* intres,
Non tangas Aures, nec barbam radere debes.

Each of the signs is described in similar triplets, from the grammar of which I conclude that *Aries* is here put for *in Ariete,* in the first hexameter.

copies, which merely repeats section 1, and is clearly spurious. It does not appear at all in the best MSS. I quote it here from MS. L.

"To knowe the degre of thyn sonne in thyn zodiak by the days in the baksyde off the Astrolabye.

[T]hanne iff þou wylte wete thatt / rekyn & knowe / qwych is the day ˙off the monyth thatt thow arte ynne, & ley thy rewle of thy astrolabye, that is to sey, the allydatha, vpon þe day in the kalendre off the Astrolabye, & he schall schewe the thy degree of the sonne."

26. After "assendent," the following additional paragraph occurs in MS. Bodley 619 ; fol. 21. It is worthy of notice, because the original of it appears in Messahala's treatise, with the title "De noticia stellarum incognitarum positarum in astrolabio." The paragraph runs thus :—

"Nota. þat by þis conclusioun þou may knowe also where ben at þat same tyme alle oþir sterres fixed þat ben sette in thin Astre-labie, and in what place of þe firmament ; And also her arising in thy orizonte, and how longe þat thei wol ben aboue þe erthe wiþ þe Arke of þe nyght / And loke euermore hov many degrees þou fynde eny sterre at þat tyme sitting vpon þin Almycanteras, and vp-on as many degrees sette þou þe reule vpon þe altitude in þe bordere ; And by the mediacioun of þy eye through þe .2. smale holes shalt thou se þe same sterre by the same altitude aforseid, And so by this conclusioun may þou redely knowe whiche is oo sterre from a-noþer in the firmament / for as many as ben in the Astrelabie. For by þat same altitude shal thou se that same sterre, & non othir / for þere ne wolle non othir alti-tude accorde þerto."

29. *Alhabor ;* i. e. Sirius or the Dog-star, as is evident from the fact of its being represented by a dog's head on the Astrolabe ; see also the table of stars on the Astrolabe, which give the declination 15° S, the latitude 39° S, and place the star in Cancer. It is also plainly described in the same table as being "in ore canis," so that it is difficult to resist the conclusion of the identity of Alhabor and Sirius. Mr Brae, following later copies that have different readings of the numbers employed, identi-fies Alhabor with Rigel or β Orionis. This is impossible, from the fact that Rigel and Alhabor *both* occur in the diagrams and tables ; see, for instance, fig. 2. It is true that Rigel was sometimes called *Algebar,* but *Alhabor* stands rather for the Arabic *El-abúr.* The Arabic name for the constellation Canis Major was *El-khelb el-akhber,* "greater dog," as distinguished from *El-khelb el-esger,* or "lesser dog ;" and the star α Canis Majoris was called *El-schira el-abûr,* from the former of which terms (*el-schira*) we get *Sirius,* and from the latter (*el-abûr*) we have *Alhabor.* See Ideler, über den Ursprung und die Bedeutung der Stern-namen, pp. 237, 256.

§ 4. "The houses [in astrology] have different powers. The strongest of all these is the first, which contains the part of the heaven about to rise : this is called the *ascendant ;* and the point of the ecliptic which is just rising is called the horoscope."—English Encyclopædia ; art. Astrology.

20. In the English Cyclopædia, art. Astrology, a quotation is given from an astrological work, in reply to the question whether the "querent" should succeed as a cattle-dealer. It contains some words very similar to Chaucer's. "If the lord of the sixth be in quartile, or in opposition to the dispositor of the part of Fortune, or the Moon, the querent cannot thrive by dealing in small cattle. The same if the lord of the sixth be afflicted either by Saturn, Mars, or the Dragon's Tail; or be found either retrograde, combust, cadent, or peregrine. [See l. 31.] The Dragon's Tail and Mars shew much loss therein by knaves and thieves, and ill bargains, &c.; and Saturn denotes much damage by the rot or murrain." The evil influence of the Dragon's Tail is treated of in the last chapter of "Hermetis Philosophi de revolutionibus nativitatum," fol. Basileæ; n. d.

30. "May see the ascendant." Cf. "Cum dominator ascendens viderit, res quæ occulta est secundum ascendentis naturam erit; quod si non videt, illud erit secundum naturam loci in quo ipse est dominator;" Cl. Ptolemæi *Centiloquium;* sect. 90.

38. *Face.* See note to Part I. sect. 21. l. 48. The late copies are very incorrect hereabouts.

§ 6. 9. Mr Brae well calls attention here to the absurd errors in the printed copies. Thynne has "in the 320 signe," and Speght "in the xxiii signe." The signs of the zodiac are only twelve, and the one opposite to the 1st is the 7th.

§ 8. I see no reason for supposing this proposition to be an interpolation, as Mr Brae suggests. Though *similar* to § 11, it is not *identical* with it. Moreover, it occurs in Messahala.

§ 9. 1. *the chapitre beforn,* i. e. a previous chapter, viz. in sect. 6. The expression supplies no argument for altering the order of the "conclusions."

4. *same manere,* i. e. a like manner. The "vulgar night" clearly means that the quantity of the "crepuscules" must be *subtracted* from the "arch of the night."

§ 13. 5. *cours,* course; *heiest cours,* highest point of the path. Late copies have *lyne;* for which Mr Brae suggested *degre.*

§ 14. 6. *but 2 degrees.* Suppose the sun's midday altitude is 49°, in latitude 52°. Then the co-latitude is 38°, and the sun's declination 11° North. This corresponds nearly (roughly speaking) to the 1st degrees of Taurus and Virgo. Which is right can "lightly" be known by the time of year, for the sun cannot be in Virgo, if the month be April. Compare sect. 15.

§ 17. This conclusion, as pointed out in the footnote, is not correct in theory, but can be made nearly so in practice, by taking the two altitudes *very near* the meridian. This is directly implied in the words "passeth any-thing the sowth westward," i. e. passes *ever so little* westward of the south line; cf. note below to 38. 9. Consequently, the first observation must also be taken very near the meridian.

24. *site,* situation. Late copies, *sight.*

§ **18**. Instead of reckoning a star's right ascension by referring it to the equator, it was reckoned by observing the degree of the zodiac which southed along with it. This is expressed in the first "Table of fixed stars" (in the Preface) by the phrase "cum gradibus, quibus celum mediant;" the other co-ordinate of position was the star's declination from the equator, as in the modern method. The ancients also used the co-ordinates of longitude and latitude of a star, the longitude being reckoned along the ecliptic, and the latitude along great circles through the poles of the ecliptic; as appears from the second Table in the Preface.

§ **19**. 6. *equinoxial*. This, as explained in the footnote, should be "ecliptik;" but I can find no MS. authority for the alteration.

§ **22**. 13. *place*. Late copies *planet;* absurdly. Latitudes of several places are given in old Latin MSS. They are frequently incorrect.

§ **23**. 3. The star A is shewn by the numbers to be the Pole-star, and is obviously the one to be observed in order to find the altitude of the Pole. What the star F is, is of no consequence. The numbers used in other copies are different, and much less satisfactory. That the star A is the Pole-star or some star near the pole in this "conclusion," is rendered probable also by the wording of the next "conclusion;" which extends the working of it to the case of any other star, provided it be a star that never sets.

§ **25**. 18. When Chaucer says that the latitude of Oxford is "certain minutes less," he probably means no more than that the latitude of Oxford was 51 degrees and 50 minutes, as in the text. For I suspect the original reading of the passage made the sun's altitude 38 degrees only, and the latitude 52 degrees; indeed, the passage stands so in MSS. C and P, both good authorities. But he added the statement that the latitude of Oxford was less than 52 degrees. It is probable that, on second thoughts, he put in the number of *minutes*, and forgot to strike out the clause " I sey nat this," &c., which was no longer necessary. Minutes were seldom reckoned otherwise than by *tens;* "a few minutes less than 50" (say 47) is a refinement to which the ancients seldom attained. Hence the amount of 10 minutes is vaguely spoken of in l. 30 as "odde Minutes." Minutes were clearly not much considered. In the present case, we are assisted by Chaucer's express statement in sect. 22. l. 6.

§ **26**. 8—11. It is singular that this sentence, obviously wanted, should appear only in one MS., and has, accordingly, been deficient in all previous editions. There can be no doubt about the genuineness of it, as it so exactly gives the right sense, and happily supplies the words "right orisonte" in l. 11 ; thus enabling the author to say, as in l. 20 he *does* say—" this *forseid* rihte orisonte."

15. *this figure*. Here occurs, in some of the MSS., a diagram representing a circle, i. e. a disc of the astrolabe, with straight lines drawn across it from left to right.

16. *Assensiouns in the rihte cercle*. This exactly answers to our modern "right ascension." We hence obtain the true origin of the

phrase. "Right ascension" was, originally, the ascension of stars at places situate *on the equator*, and was most conveniently measured along the equatorial circle, by observation of the times of transit of the various stars across the meridian. In other latitudes, the ascension of every degree of the *zodiac* could be easily tabulated by observing what degree of the equator came to the meridian with the said degree of the zodiac ; see l. 18. It hence appears that, whilst persisting in using "longitudes" and reckoning along the zodiac, the ancients were obliged, in practice, to refer the degrees of longitude to the equator. The modern method of recognizing this necessity, and registering right ascensions as of more importance than longitudes, is a great improvement. The ancients were restrained from it by their unnecessary reverence for the zodiac. Cf. Ptolemy's Almagest, lib. i. c. xiii.

§ 29. Chaucer omits to say that the experiment should be made when the sun is very nearly on the meridian. Otherwise, the confusion of the azimuth with the hour-angle might cause a considerable error.

§ 30. That the phrase "wey of the sonne" really means the sun's apparent *diurnal* course in this conclusion, may be further seen by consulting the Latin of Messahala. Mr Brae objects to this; but I see no limit to the planet's position in the words of the conclusion. Cf. the Critical Note.

§ 31. In my footnote, I have used the expression " it does not mean, *as it should*, the zenith point." I mean—"as, according to our modern ideas, it should ; "—for the derivation of *zenith* shews that the meaning used in this proposition is the older meaning of the two. See note above to i. 19. 7.

5. 24 *parties*. These 24 parts were suggested by the 24 hours of the day. The "32 parts" used by "shipmen" are due to the continual halving of angles. Thus, the four cardinal points have points half-way between them, making eight points; between which, we can insert eight more, making sixteen ; and between these, sixteen more, making thirty-two. Hence the 32 points of the compass.

§ 33. 5. We should probably insert *or sowth* after the word *north*. Cf. the Critical Note.

§ 34. 3. That "vpon the mones side" means nearly in the same azimuth as the moon is apparent from l. 11 below, where Chaucer says that some treatises make no exception even if the star is *not* quite in the same azimuth. This was certainly a rough mode of observation.

§ 35. 9. *riht side*, East side. See i. 6. 1.

18. *Episicle*, epicycle. To account for the planetary motions, epicycles were invented. The moon, for instance, was supposed to revolve round a *moving* centre, which centre itself moved round the earth in a perfect circle. This came a little nearer to the true motion in some instances, but was hopelessly wrong, and nothing could be made of it, even when a *second* epicycle, revolving about a centre which moved in the *first* epicycle, was superadded. All that Chaucer says here is, that, whilst the centre of the moon's epicycle had a direct

motion, the moon's motion in the epicycle itself was a reverse one, unlike that of the other planetary bodies. The subject is hardly worth further discussion, so I merely refer the reader to the Almagest, lib. iv. c. 5; and lib. ix. c. 5.

§ 36. The "equations of houses" means the dividing of the sphere into *equal* portions, and the right numbering of those portions or houses. The most important house was the first, or ascendent, just rising; the next in importance was the tenth, which was just coming on the meridian; then come the seventh or descendent, just about to set, and the fourth, just coming to the line of midnight. The next in importance were the *succedents*, or houses immediately following these, viz. the second, the eleventh, the eighth, and the fifth. See *Succedent* in the Glossary.

§ 37. 17. *thise* 3 *howsez*. That is, the nadirs of the 2nd, 3rd, and 4th houses give the houses that "follow," i. e. the 8th, 9th, and 10th. The word "follow" here seems to refer, not to position, but to the order in which the houses may most conveniently be found. Chaucer omits to add that the beginnings of the 5th and 6th houses can be found in a similar way, because it is sufficiently evident. See the original in Messahala.

§ 38. 1. *for warpyng, the brodere the bettre.* This may mean, either (1) to prevent warpyng, the thicker the better; or (2) to prevent the errors arising from warping (because of warping) the larger the better. I believe the latter to be the true interpretation; for it is better thus to guard against possible errors than to make the plate very thick and, at the same time, small. Besides which the usual meaning of *brodere* is *wider, larger, more ample.* Indeed, we find the very expression "non sit tamen nimis parvus" in the 4th section of the *Practica Chilindri* of John Hoveden, published by the Chaucer Society; which see.

7. *fro the centre*, i. e. sticking up above the centre, the length of the wire being equal to a fourth of the diameter, or half the radius, of the circle. This proportion would do for many days in the year; but in the summer time, the pin would bear to be rather longer. Still, we need not alter the text. Cf. the Critical Note.

9. *any-thyng*, i. e. ever so little; so *ony-thyng* in l. 12.

§ 39. Though MS. A. is rather corrupt here, there is little doubt about the corrections to be made. See the Critical Notes.

19. That is, the latitude, or breadth, of a climate, or belt, is measured along a line which goes from North to South as far as the earth extends; so that the latitude of the *first* climate, for example, is measured from the beginning of it to the end of the same, in a due northerly direction. Other authors, he explains, reckoned the latitude of a climate always from the equinoxial line, instead of from the parallel of latitude which terminated the climate immediately to the south of it. Thus the latitude of the fourth climate might mean, either the breadth of that belt *itself*, or the *whole* breadth from the equator to the Northern limit of that climate. The MS. E. 2 in St John's College, Cambridge,

contains (besides Chaucer's " Astrolabe ") a Latin treatise entitled " De
septem climatibus expositio." We find mention of the " climates " also
in MS. Camb. Ii. 3. 3, fol. 33 *b*, where a diagram appears representing
a hemisphere, divided by parallels of latitude into 9 climates or belts,
which, beginning from the equator, are as follows. 1. Inhabitabile
propter Calorem. 2. Primum clima dia Meroes. 3. Secundum clima
dya cienes. 4. Tertium clima di' alexandrios. 5. Quartum clima dia
rodos. 6. Quintum clima dia romes. 7. Sextum clima dia boristenes.
8. Septimum clima dia rifeos. 9. Inhabitabile. This agrees with the
list in the footnote on p. 48.

There is a passage in Mandeville which well illustrates Chaucer ; I
quote the part of it which more· immediately relates to the Climates.
"For the Superficialtee of the Erthe is departed in 7 parties, for the
7 Planetes ; and the parties ben clept Clymates. And our parties be
not of the 7 Clymates : for they ben descendynge toward the West.
And also these yles of Ynde, which beth evene aȝenst us, beth noght
reckned in the Climates : for thei ben aȝenst us, that ben in the lowe
Contree. And the 7 Clymates strecchen hem, envyrounynge the World,"
&c. Mandeville's Voiage, ed. Halliwell, p. 186. See also Ptolemy's
Almagest, lib. ii.

As regards the longitudes of towns, it may be observed that in MS.
F. 25 in St John's College, Cambridge, the longitudes of Rome, Cordova,
London, Paris, and Malta, are said to. be 34°. 24', 9°. 30', 19°, 20°, and
38° respectively. These do not well agree together, but they suggest a
reckoning from a meridian situated some 20° W. from that of Greenwich.
Chaucer says nothing as to what meridian was used for reckoning longitudes from ; and Messahala is but vague.

§ 40. It is possible that this conclusion was really intended to belong
to the fourth part of the treatise, and was written by way of instalment.
See the Prologue, ll. 63—67. It is curious that in all the best MSS.
(P excepted) the last sentence should be incomplete.

12. This sentence is very awkward. It seems to mean—" and then
set I the point of F upward in the same sign, because that the latitude
was north, upon the latitude of Venus ; that is to say, (I set it upward)
keeping it in the 6th degree of Capricorn." *Upward* means inward, i. e.
towards the centre or towards the north ; the opposite being expressed
by southward, or outward, or toward the border, as in l. 46 below. *Upon
the latitude of Venus* means that the point F of the compass was set upon
the second degree of latitude, so that the space between the legs of the
compass became equal to 2 degrees, as said in l. 15. Lastly, the words
that is to seyn, in the 6 degree, &c., are an explanation of the vaguer expression *in the same signe*. The repetition of the words *that is to seyn*,
&c. (ll. 11 and 14), is intended to draw attention to the necessity of
keeping *both* legs of the compass in the same degree of longitude.

55. Possibly Chaucer left the sentence incomplete. The words
" thou shalt do well enough " may easily have been added by another
hand to bring the sentence to an apparent, though not wholly satisfac-

tory, conclusion. *Colophon.* This colophon is written (in a later hand) in MS. A at the bottom of the page, a part of which, after the words "howre after howre," is left blank.

41—43. I have mended the text as well as I could by words, &c., inserted between square brackets. Nearly all the emendations rest on authority; see the Critical Notes. The text is not a good one, but I do not see why these sections may not have been written by Chaucer. For a definition of the terms "Umbra Extensa" and "Umbra Versa," see sections 5 and 6 of the *Practica Chilindri* of John Hoveden, published by the Chaucer Society. The *umbra extensa* or *recta* is the shadow cast on a plain by any perfectly upright object; but the restriction is commonly introduced, that the altitude of the sun shall exceed 45°. The *umbra versa* is the shadow cast *perpendicularly* downwards along a wall by a style which projects from the wall at right angles to it; the restriction is commonly introduced, that the sun's altitude shall be less than 45°. The *umbra versa* is the one which appeared on the "chylindre"; hence John de Hoveden explains how to calculate the altitude of an object by it.

44. This article and the next may possibly be Chaucer's. It is well-known that he speaks of "collect" and "expans yeres" and "rotes" in the Frankeleines Tale; Cant. Ta. l. 11587; the note upon which in the glossary to Urry's Chaucer may be found also in Tyrwhitt's Glossary, s. v. *Expans;* but it is worth while to repeat it here. "In this and the following verses, the Poet describes the Alphonsine Astronomical Tables by the several parts of them, wherein some technical terms occur, which were used by the old astronomers, and continued by the compilers of those tables. *Collect* years are certain sums of years, with the motions of the heavenly bodies corresponding to them, as of 20, 40, 60, &c., disposed into tables; and *Expans* years are the single years, with the motions of the heavenly bodies answering to them, beginning at 1, and continued on to the smallest *Collect* sum, as 20. A *Root*, or *Radix*, is any certain time taken at pleasure, from which, as an era, the celestial motions are to be computed. By 'proporcionel convenientes' [C. T. l. 11590] are meant the Tables of Proportional parts." To which Tyrwhitt adds, from Chambers's Encyclopædia, with reference to C. T. l. 11589, that "*Argument* in astronomy is an arc whereby we seek another unknown arc proportional to [or rather, dependent upon] the first." See also *Entere* in the Glossary.

Tables of mean motions of the Sun are given in Ptolemy's Almagest, lib. iii. c. 2; of the Moon, lib. iv. c. 3; of the Planets, lib. viii. c. 3; also in MS. Ii. 3. 3, fol. 88*b*, &c.

41*a*—42*b*. The fact that these articles are mere repetitions of sections 41—43 is almost conclusive against their genuineness. I do not suppose that sect. 46 is Chaucer's either, but it is added for the sake of completeness.

𝔄𝔰𝔱𝔯𝔬𝔩𝔞𝔟𝔦𝔲𝔪 𝔐𝔢𝔰𝔰𝔢𝔥𝔞𝔩𝔩𝔢.

[*MS. Camb. Univ. Lib. Ii. 3. 3, p. 74.*]

Nomina instrumentorum sunt hec. [1]Primum est armilla suspensoria ad capiendam altitudinem, et dicitur arabice alhahucia. [2]Secundum est alhabor, id est, ansa que iungitur ei. [3]Postea mater, 4 rotula scilicet, in se continens omnes tabulas cum aranea, cui coniungitur margolabrum scilicet in .360. gradus diuisum. [4]Tabule autem ab hac contente figurantur tribus circulis quorum minor est circulus cancri, et medius est circulus equinoctialis, et maximus 8 circulus capricorni. [5]Postea circulus almucantherath, qui sunt circuli in medietate superiori descripti quorum quidam sunt integri, quidam apparent imperfecti ; quibus prior est orizon, et diuidit duo emisperia. Centrum autem interioris almucantherath cenit capitum 12 nominatur. [6]Deinde est azimuth, qui sunt partes circulorum almucantherath intersecantes. [7]Post quas sunt hore, in medietate inferiori descripte. In[ter] horas .2. sunt crepusculorum lince. [8]Postea linea medii celi, que est linea descendens ab armilla per centrum in 16 oppositam partem astrolabii, cuius medietas a centro in armillam dicitur linea meridiei ; et alia dicitur angulus terre et medie noctis. [9]Post hec et sequitur alhanthabuth, id est aranea, in quo sunt signa cum zodiaco constituta, stelle quoque fixe, in quo via dicitur esse 20 solis ; et quicquid fuerit infra motum capitis arietis et libre, ex hoc zodiaco, septemtrionale ; quod autem extra, meridianum. [10]Sequitur

---

[1] Part I. § 1.   [2] Pt I. § 2.   [3] Pt I. §§ 3, 16.   [4] Pt I. § 17.
[5] Pt I. § 18.   [6] Pt I. § 19.   [7] Pt I. § 20.   [8] Cf. Pt I. §§ 4, 15.
[9] Pt I. § 21.   [10] Pt I. § 23.

almuri, quod ostensor dicitur latine, denticulus scilicet, extra circulum capricorni; in alhanthabuth relictus deinde almenath, id est,
foramen quod est in medio rethis; [1]in quo est axis retinens tabulas 24
climatum, in quam intrat alphaeraz, id est, equus restringens araneam
cum rotula, quasi cuneus. Et in illa parte matris sunt .2. circuli equa-
tionis solis exterius, quorum [2]unus continet numerum dierum anni
.365., [3]et scribentur sub eo nomina mensium. [4]Et alius signorum 28
gradus, [5]et infra eum scribuntur nomina signorum. [6]Postea quarta
capiende altitudinis. Postea quadrans, cuius latera in .12. puncta
diuisa sunt. [7]Sequitur regula, que circumuoluitur in dorso astrolabii, in qua sunt tabule perforate, ad capiendum altitudinem solis 32
in die, stellarum in nocte.

[8]Cum uolueris scire gradum solis, pone regulam super diem
mensis presentis, et gradus a summitate eius tactus erit gradus solis,
qui cuius signi sit uidebis, et eum ex alia parte nota in zodiaco in 36
rethi. Notabis et nadayz eius, quod est simul gradus .7. signi. Diem
quoque mensis per gradum solis inuenies; posita enim regula super
gradum solis diem quesitum ostendet.

[9]De altitudine solis et stellarum inuenienda capitulum.   40

Cum vis altitudinem solis scire, Suspende astrolabium de manu
tua dextra per eius armillam, et sinistro tuo latere soli opposito,
subleua vel deprime regulam, donec radius solis per utriusque tabule
foramen transeat; quo facto, vide quot gradus a linea orientali 44
eleuatur regula, et est solis altitudo; similiter facies in nocte, per
stellas fixas.

[10]Si autem vis scire certitudinem hore et etiam ascendentis, pone
gradum solis super almucantherath altitudinis, ex parte orientis, si 48
altitudo sit ante meridiem; aut ex parte occidentis, si altitudo sit
accepta post medium diem; et super quam horam ceciderit nadayz
gradus solis erit hora presens, et signum quod fuit ex parte orizontis
orientalis, est oriens, id est, ascendens; quod uero in occidentali, 52

---

[1] Pt I. § 14.   [2] Pt I. § 9.   [3] Pt I. § 10.   [4] Pt I. § 7.
[5] Pt I. § 8.   [6] Pt I. § 12.   [7] Pt I. § 13.   [8] Pt II. § 1.
[9] Pt II. § 2.   [10] Pt II. § 3.

occidens. Quod uero ceciderit in linea medii celi est in medio celo,
et eius nadays angulus terre.

[1]Et si ceciderit inter duo almuchantherath, vide differentiam
56 numeri inter almuchantherath precedentem et altitudinem solis, et
denomina differentiam de numero longitudinis almuchantherath, quod
est .6., si almuchantherat continet .6. gradus et .6.; quod si almuchantherath contineat .3. gradus et .3., denomina partem illorum de .3.; et
60 sic de aliis. Postea scito motum almuri ab initio primi almuchanthanth usque ad inicium secundi de gradibus marginis; et pone
super illorum partem denominatam ab eis, secundum proportionem
differentie dicte, ex .6. vel de .3. gradibus; et tunc habebis certum
64 gradum inter duo almuchantherath; et tunc considera eas horas, &c.,
sicut dictum est superius. Si illud idem in nocte scire desideras,
accipe altitudinem alicuius stelle, in alhanthabuth descripte, que
transit ex parte orientis uel occidentis; et pone cacumen illius stelle
68 in almucantherath sue altitudinis, et gradus solis indicabit tibi horas
noctis, sicut nadayz eius diei; de aliis fac omnibus sicut dictum est
in superioribus.

De crepusculo vespertino et matutino.

72   Cum uolueris scire finem crepusculi uespertini et inicium
matutini, vide cum uenerit gradus solis ad lineam crepusculi occidentalis; tunc est finis eius; et cum ad orientalem, est inicium
crepusculi.

76   [2]**Aliter idem.**

Uel sic; vide quum nadayz solis uenerit ad .18. gradum
almuchantherath in oriente, erit finis crepusculi uespertini; et cum
venerit ad .18. gradum almuchanthanth in occidente, est initium
80 crepusculi matutini; et hec est leuis.

[3]De inuencione arcus diurni et nocturni: Rubrica.

Si vis scire arcum diei et noctis, pone locum solis, id est, gradum
in quo est super primum almucantherath; et nota locum almuri
84 inter gradus limbi; post hec moue gradum solis usque ad occidentem;

---

[1] Pt II. § 5.   [2] Pt II. § 6.   [3] Pt II. § 7.

et nota etiam locum eiusdem in ipsis gradibus ; et motus eius ab una nota in aliam est arcus diei ; reliqua uero pars circuli est arcus noctis, quia illa duo continebunt .360. gradus, que est quantitas diei et noctis ; similiter facies de stellis fixis, si uolueris scire earum 88 moram super terram.

1De quantitate horarum diei inequalium.

Si volueris quantitatem horarum inequalium diei scire, diuide arcum diei per .12., et habebis numerum graduum hore diurne ; 92 quem si subtrahis a .30. remanebit numerus graduum hore nocturne, quia hora inequalis nocturna cum hora inequali diurna facit .30. gradus in omni die, qui sunt due hore equales.

Si horas diei uolueris querere equales, diuide arcum diei per .15., 96 et habebis numerum horarum equalium ; similiter in nocte.

De parte hore partita inuenienda per almuri capitulum.

Cum transierit pars hore, et uolueris scire quota pars sit hore, scito numerum graduum in labro ab inicio hore illius in almuri ; et 100 quomodo ille numerus se habebit ad numerum totius hore, sic pars transacta se habebit ad totam horam.

2De numero horarum diei equalium preteritarum inueniendo capitulum. 104

Si uolueris scire quot hore equales transierunt de die, accipe gradum solis, et pone super almuchanthanth altitudinis et signa locum almuri in gradibus. Postea uolue retro gradum solis usque ad primum gradum almuchantherath in oriente ; et secundo nota eius-108 dem locum ; post hec diuide gradus qui sunt inter .2. notas per .15., et habebis horas equales. Similiter facies de nocte ; postquam enim inueneris horam inequalem per gradum et altitudinem alicuius stelle, signato loco almuri, reduces gradum solis ad orizontem occidentalem, 112 et notabis iterum locum almuri. Spacium inter hec duo loca diuides, sicut prius, per .15. scilicet, et inuenies. Eodem modo scies quot sint hore equales inter meridiem et quemlibet punctum alium, et quodlibet instans. 116

---

1 Pt II. § 10.   2 Pt II. § 11.

[1]De conuersione horarum inequalium in horas equales : Rubrica.

Si volueris reducere horas inequales in horas equales, scito gradus horarum inequalium, quot sint; et diuide eos per .15., et 120 habebis horas equales; similiter facies de horis equalibus.

[2]De altitudine solis in meridie habenda capitulum.

Si uolueris scire altitudinem solis in media die, quod est inicium recessionis, pone gradum solis super lineam medii celi; et numerus 124 graduum almucantherath a loco solis in orizontem est altitudo eiusdem medie diei. Similiter fac cum stellis fixis.

Inuencio hore diei per allidadam : capitulum.

Si per allidadam horariam uis scire horam diei naturalem, pone 128 allidadam super altitudinem medie diei illius in dorso astrolabii suspensi; et uerte dorsum ad solem tam diu donec umbra vniuscuiusque anguli superioris pinnule ˍcadat in allidada, quelibet in directo sui lateris; et vbi occiderit in diuisionibus erit hora quesita.

132 De eodem inueniendo per lineas.

Item per allidadam in dorso et lineas horarum inter latera gnomonis, si sint posite ut in quadrante, sic. Super altitudinem solis meridianam in illa die pone allidadam; et nota vbi meridianus 136 circulus, id est, linea finis .6. hore, secuerit lineam fiducie ipsius allidade; et pone ibi signum de incausto; et illud signum valet situationem margarite in quadrante; deinde accipe altitudinem solis in quacunque hora vis, et illud signum inter horas dabit horam 140 naturalem, ut in quadrante.

[3]Capitulum preambulum ad quedam sequencia.

Amplius scito quod circulus signorum diuiditur in .2. semicirculos, quorum vnus est a capite capricorni in caput cancri, et alius 144 a capite cancri in caput capricorni; et caput capricorni est solsticium

---

[1] Pt II. § 8. [2] Pt II. § 13. [3] Pt II. § 16.

hyemale, caput cancri estiuale. Scito et quod omnis duo equidis-
tantes gradus ab aliquo horum solsticiorum sunt vnius declinacionis
versus septentrionalem vel meridiem; et dies eorum vel noctes sunt
equales, et umbre et altitudines in media die sunt equales. **148**

[1]De gradu solis ignoto per rethe habendo.

Si volueris cognoscere gradum solis ignotum, pone notam super
altitudinem medie diei, quam sumpsisti prius per regulam in dorso
astrolabii; deinde uolue rethe, cadentque duo gradus super ipsam 152
notam; quorum vnum scies esse gradum solis per signum mensis
cuius fuerit dies.

[2]Quis dies cui diei sit equalis.

Si volueris scire que dies cui diei sit equalis, scies hoc per 156
gradum equedistantem a solsticiis, quia eorum dies sunt equales,
sicut dictum est superius.

[3]De Inuencione gradus stelle cum quo celum mediat.

Si uolueris scire cum quo gradu uenit stella aliqua ad medium 160
diem, uel oritur; pone stellam super lineam medie diei, quia gradus
qui cecidit super eandem lineam est gradus quesitus; similiter fac ad
lineam orientalem et occidentalem. Gradum uero longitudinis
habebis per filum positum super polum zodiaci, per totam declina- 164
cionem inuentum.

[4]De altitudine cenith solis habendi.

Si uolueris cenith solis scire, accipe altitudinem eius hora qua
uolueris hoc scire, et pone gradum solis super almucanthanth alti- 168
tudinis in parte qua fuerit, sicut facis ad inuentionem horarum.
Post hec, accipe quid congruit gradui solis de azimuth, et super quem
gradum sit cenith de quarta que opponitur; et necesse est ut hec
quarta sit meridiana orientalis, uel septentrionalis orientalis; aut 172
occidentalis meridiana, uel septentrionalis occidentalis; et similiter
facies de stellis fixis per earum altitudines.

---

[1] Pt II. § 14.   [2] Pt II. § 15.   [3] Cf. Pt II. § 18.   [4] Pt II. § 33.

[1]De cenith ortus solis habendo, et aliorum planetarum.

176   Et si uolueris scire cenith ortus solis, vel alicuius stelle fixe, pone gradum solis uel stellam super orizontem orientalem, et aspice quid sibi accidat de azimuth, similiter quam sit ortus; et hoc est cenith ortus, et super simile eius erit occasus in simili eius quarta, siue 180 orientalis, siue meridionalis fuerit.

[2]De quatuor plagis mundi: Rubrica.

Ad habendas quatuor plagas mundi veraciter, accipe altitudinem solis ut supra, et vide in quota quarta sit; deinde vide in qua 184 altitudine ipse gradus solis sit inter lineas azimuth in principio quarte orientalis, que incipit a coluro septentrionali siue a medie noctis linea, a qua incipies computare; et quotus fuerit numerus, tantum sume in dorso astrolabii, ab ipso coluro uersus armillam, pro- 188 cedendum per orizontem, si est ante meridiem, vel per occidentem, si est post meridiem; et vbi numerus idem finitur, ibi pone regulam; deinde astrolabium utraque manu tenens, sursum uersa eius posteriori superficie, diligenter te oppone soli, donec radius solis transeat per 192 ambo foramina; tunc caute illud pone super terram, ut non moueatur ad aliquam partem; habebis quatuor lineas in centro astrolabii con- currentes, quatuor mundi plagas directe oppositas indicantes, scilicet orientalem, occidentalem, &c.; similiter operabis in nocte per 196 stellam fixam. Vel locata iam regula in dorso astrolabii, sursum uersa eius facie, eque distanter orizonti ut proximo dictum est, fac umbram amborum angulorum pinnule cadere super .2. latera regule, scilicet, dextram umbram super latus dextrum, et sinistram umbram 200 super sinistrum latus; et statim habebis quatuor lineas et quatuor plagas mundi predictas.

[3]De declinacione cuiuslibet gradus habenda.

Si scire uolueris declinacionem cuiuslibet gradus signorum, pone 204 super lineam medii celi uel diei, et scito eius altitudinem ab oriente; postea scito altitudinem capitis arietis et libre in eadem linea; deinde scito altitudinem utramque, et differentia ipsarum altitudinum

---

[1] Cf. Pt II. § 31.   [2] Pt II. § 29.   [3] Pt II. § 20.

est declinatio eiusdem gradus ab equinoctiali linea. Si autem
gradus signi fuerit septentrionalis, est declinacio septentrionalis; si 208
meridiana, meridiana. Scito etiam quod gradus signorum septen-
trionalium sunt altiores equinoctio, quod est in capite arietis et eius
opposito; et meridionalium inferiores, secundum declinationes eorum
ab eo. Maior autem declinacio est in capite cancri et capricorni: 212
eodem modo inuenies declinacionem stellarum fixarum.

[1]De altitudine poli vel latitudine regionis.

Scito quod altitudo regionis sit latitudo cenith capitum[2] ab
equinoctiali circulo uersus septentrionalem vel meridiem, que similis 216
est altitudini poli septentrionalis, et depressioni eius oppositi ab
orizonte, que duo sunt in parte equales. Cum ergo latitudinem
cuiusque regionis scire uolueris, altitudinem solis in media die considera, quam minues de .90., si fuerit sol in inicio arietis et libre, et 220
quod est residuum erit latitudo regionis; tunc enim motus solis erit
in equinoctiali linea. Si uero in alio gradu fuerit sol, eiusdem
gradus declinacionem considera per tabulam declinacionis solis, uel
per regulas ante datas; quam minues de altitudine solis in medio 224
die, si fuerit septentrionalis; si uero meridionalis, adde illam; et
habebis altitudinem inicii arietis in regione illa, quam subtrahes
sicut predictum est a .90., et quod remanserit est distancia regionis
ab equinoctiali linea. 228

De eodem, scilicet aliter, capitulum: Rubrica.

Uel si volueris accipere altitudinem cuiusuis stelle altiorem, et
eius elongacionem ab equinoctiali linea considera; cum qua fac ut
supra dictum est. [3]Vel quere cuiusuis stelle non occidentis in eadem 232
regione altitudinem altiorem et inferiorem, et utriusque insimiliter
collecte tolle medietatem, que est altitudo poli in eadem regione.

[4]De noticia tabule almucanterat.

Si uis scire ad quam latitudinem facta sit tabula almucantheralis, 236
vide in linea meridiana quot almucantherath sint in circulo equi-

---

[1] Pt II. § 25. [2] cunū eius (sic); MS. Hh. 6. 8 has capitum; see l. 11.
[3] Pt II. § 24. [4] Pt II. § 21.

noctiali usque ad cenith, vel ab axe ad orizontem in septentrione; et
super tantam latitudinem facta est tabula : altitudo uero arietis est
240 tot graduum quot fuerint ab eodem circulo ad orizontem, vel a cenith
ad axem.

De horis inueniendis per tabulas latitudinis : Rubrica.

Cvm in aliquo regione, cuius latitudo in tabulis astrolabii non
244 fuerit descripta, uolueris inuenire per illud astrolabium horas illas,
regionis latitudinis et latitudinis maioris propinquioris sibi et
minoris ibi descripte nota differentiam; deinde proportionem illius
differentie ad differentiam que est inter minorem latitudinem ibi
248 descriptam et maiorem, inter quas videlicet est latitudo regionis
illius, memorie commenda. Postea uero accepta solis altitudine in
eadem regione, quere horas per latitudinem minorem, et similiter per
latitudinem maiorem, et harum horarum diuersarum differentie tolle
252 partem proportionalem, secundum proportionem differentic superius
sumptam; quam partem addes horis minoris latitudinis, si fuerint
pauciores horis maioris latitudinis, vel subtrahes ab eisdem, si fuerint
plures; et que tunc remanserint erunt hore illius regionis; similiter
256 facies in horis noctis et in aliis operibus.

[1] De gradu solis ignoto habendo.

Cvm qualibet die gradum solis per alhanthabuth uolueris
inuenire, altitudinem eius in media die considera, quam notabis in
260 almucanthanth in meridiana linea; tunc quartam circuli signorum in
qua fuerit sol gira;[2] et gradus qui continget notam altitudinis in
media linea est gradus solis.

De longitudine inter duas regiones habenda per eclypsim.

264   [3] Longitudo regionis ab alia est distancia meridiani circuli vnius
a meridiano circulo alterius. Cumque uolueris scire longitudinem
inter .2. regiones, considera inicium eclypsis lunaris, per quot horas
equales distet a medio precedentis diei in utrisque regionibus.
268 Deinde minue horas vnius regionis de horis alterius, et que reman-

---

[1] Pt II. § 14.   [2] *Written* giza ; *but* MS. Hh. 6. 8 *has* gira.
[3] Cf. Pt II. § 39. l. 11.

serint erunt hore longitudinis inter utrasque ; multiplica ita ea in
.15., et habebis quot gradus sit earum longitudo ab inuicem.   Longitudines quarundam regionum, hoc est, elongationes circulorum earum
meridianorum a meridiano circulo ultime regionis habitabilis in 272
occidente, et earum longitudines et distancias ab equinoctiali circulo
notabimus in quadam tabula sufficienter.

De eodem in miliaribus capitulum.

  Si quot miliaria sint in[ter] .2. regiones a se inuicem distantes 276
noscere queris, longitudinem et latitudinem inter utrasque considera ;
deinde longitudinem in se ductam latitudini in se multiplicate aggrega,
et collige ; inde summe tolle radicem, et unicuique gradui ipsius radicis
et dimidio da .100. miliaria ; et per tot [miliaria] distat vna regio ab alia. 280
Si autem earum latitudo fuerit eadem, fac cum gradu longitudinis
tantum, sicut deberet fieri cum gradu radicis.   Si uero longitudo
fuerit vna, fac cum latitudine tantum, et inuenies quod queris.

[1]Scientia ascensionis signorum in circulo directo.   284

Si autem ascensiones signorum in circulo directo scire desideras,
inicium cuiusuis signi super lineam meridianam pone, et locum
almuri in margine nota ; postea moue rethe donec finis signi cadat
super lineam meridiei, et gradus quibus mouebitur almuri erunt 288
ascensiones eiusdem signi ; et similiter facies ad quamlibet portionem
circuli.

[2]De ascensionibus signorum in circulo obliquo.

Ascensiones autem signorum in qualibet regione sic poteris 292
inuenire ; moue rethe ab inicio signi usque ad finem eiusdem, et
gradus quibus mouetur in margine almuri erunt ascensiones signorum
in eadem regione ; mouebis enim signum in orizontis parte orientali,
ut scias eius ascensionem : vt autem scias eius moram in occasu, 296
mouebis illud in orizontis parte occidentali ; ita etiam fiet in qualibet
circuli portione.   Gradibus eciam ascensionum diuisis per .15., et re-
siduo pro horis fractibus (*sic*) computato, habebis horas equales, uel eis
diuisis per numerum graduum hore inequalis, patebit per quot horas 300

---

[1] Pt II. § 27.   [2] Pt II. § 28.

naturales uel inequales, cum fractionibus, quodlibet signum uel
planeta uel quelibet portio ascendat uel occidat in qualibet regione.

[1]De noticia stellarum incognitarum positarum in astrolabio.

304   Ut habeatis noticiam stellarum incognitarum que posite sunt in
astrolabio, sume primo altitudinem alicuius stelle note, et pone eam
in almucantherath, super similem altitudinem; postea vide stellam
quam uolueris scire, super quantam altitudinem iaceat inter almu-
308 cantherath, et in qua parte sit, scilicet, in oriente vel occidente; quo
uiso, pone eam in dorso astrolabii super eandem altitudinem, et verte
illud ad eandem plagam celi in qua accepisti stellam; et maior stella
quam vides per foramina regule ipsa est quam queris.

312 De noticia stellarum incognitarum non positarum in astrolabio.

Scire uolens gradum stelle ignote, in astrolabio non posite, uel
planete, expecta donec ille planeta vel stella sit in meridie; deinde
visa aliqua stella cuius locum pro certo scias et astrolabio insignite,
316 secundum altitudinem eius rethe dispone, ponendo stellam inter
almuchantherath super similem altitudinem; et directo gradus signorum qui erit in linea medii celi erit stella de qua dubitas, et est
longitudo eius nota; latitudo patet, computatis almucantherath a
320 nota illius altitudinis usque ad equinoctialem. Potes eciam per occasum solis rethe tuum disponere, si nullam stellam cognoueris, et sic
cognosces omnes stellas.

[2]Ad sciendum in quo gradu signi luna sit: Rubrica.

324   Cvm in quo gradu signi luna[3] sit scire uolueris, altitudinem lune
considera; et eam in almucantherath, in parte in qua fuerit, nota;
deinde stellam aliquam in rethi constitutam super altitudinem suam
in eadem hora cum altitudine lune acceptam, in parte qua fuerit,
328 pone; et gradus circuli zodiaci qui ceciderit inter almucantherath
super altitudinem lune, eritque gradus lune. Si autem apparuerit in
die, idem facies cum altitudine illius et altitudine solis, Considera

---

[1] See *Additional Note* to Pt II. § 3. 1. 26.   [2] Pt II. § 34.
[3] *Written* stella; *but* MS. Hh. 6. 8 *has* luna.

igitur cuius signi sit gradus. Idem poteris quoque eodem modo
planetarum loca inuestigare, si eorum altitudinem in nocte poteris 332
notare.

De loco lune inueniendo capitulum.

Cvm in quo gradu sit luna scire desideras, quot dies habeat
mensis lunaris in eadem die considera, quibus duplicatis, quod col- 336
lectum fuerit distribue per .5., dando cuilibet signo .5., et incipias a
signo in quo fuerit sol; et vbi numerus finierit, in eodem signo est
luna; et si remanserit, id est, infra .5., iam perambulauit luna .6.
gradus.   340

[1] De locis planetarum inueniendis.

Loca planetarum poteris in alio modo inuestigare, et verius.
Sume altitudinem planete quum est iuxta lineam medii celi, et serua
eam. Item, sume ad eandem horam ascendens per aliquam stellarum 344
fixarum, et hoc serua eciam cum hora; posthec vide quum ille
planeta incipiat descendere a linea medii celi, et sume eius altitudinem quum sit equalis altitudini prius sumpte ante lineam medii
celi; et iterum in eadem hora sume ascendens et horam per aliquam 348
stellam fixam; deinde sume medium inter ascendens primum et secundum per almuri in limbo; et gradus qui ceciderit tunc super
lineam medii celi, in illo est planeta.

[2] De latitudine planetarum a via solis inuenienda.   352

Scire uolens utrum planeta sit australis uel septentrionalis in via
solis, considera utrum altitudo quam sumpsisti quando erat prope
lineam medii celi sit equalis altitudini gradus in quo est planeta, vel
maior, vel minor; si enim est equalis, tunc directe est in via solis, 356
et nullam habet latitudinem; si autem altitudo planete sit maior
quam gradus in quo est sol,[3] tunc planeta est septentrionalis a uia
solis; si minor, tunc est australis; et tantum declinat a via solis
quantum est maior vel minor.   360

[4] De directione et retrogradacione planetarum.

Utrum planeta sit retrogradus uel directus sic poteris inquirere;
cuiusuis eorum altitudinem et altitudinem stelle quoque fixe memorie

---

[1] Pt II. § 17.   [2] Pt II. § 30.   [3] MS. planeta.   [4] Pt II. § 35.

364 commenda; deinde post terciam noctem uel quartam, in qua est sen-
sibilis motus, cum stelle fuerint in eadem altitudine prime altitudini
et altitudinem planete considera; quum, si fuerit minor sua altitudine prima, planeta est directus, si fuerit in parte orientali; et si
368 fuerit parte occidentali, retrogradus; et si secunda altitudo planete
fuerit maior prima, est retrogradus, si hora accepte altitudinis fuerit
ex parte orientis; et si fuerit ex parte occidentis, est directus. Oppositum autem de partibus noueris esse in luna.

372   ¹**De equacione .12. domorum per astrolabium.**

Cvm .12. domos uolueris adequare, gradum ascendentem super
lineam .8. hore pone; tunc gradus qui ceciderit super lineam medie
noctis est inicium secunde domus. Deinde reducto gradu ascendentis
376 ad finem .10. hore, gradus inuentus super predictam lineam medie
noctis est inicium .3. domus. Reduces quoque eundem gradum ad
orizontem orientalem, et erit eius nadayz in orizonte occidentis;
gradus uero in eadem prenominata linea existens erit inicium .4.
380 domus. Pones etiam nadayz gradus ascendentis super finem .2.
hore, et tunc predicta linea indicabit tibi inicium .5. domus. Si
autem posueris idem nadayz super finem .4. hore, cadet inicium .6.
domus super eandem lineam medie noctis. Inicium autem .7. domus
384 est nadayz ascendentis. Et inicium .8. nadayz secunde; principium
.9° nadayz .3°.; et .10° nadayz quarte. Principium vndecime nadayz
.5˙ et .12. nadayz sexte.

²**De eodem, scilicet aliter, capitulum.**

388   Item, habito ascendente et aliis tribus angulis, pone regulam
nouiter super rethe constitutam super gradum ascendentem, et gradus
limbi inter eam et armillam uel punctum meridianum diuisi in .3.
partes sunt ascensiones trium domorum ab ascendente in meridiem;
392 vnde si posueris eam super primam .3. ab ascendente, habebis in
zodiaco inicium .12. domus, et super secundam .3., inicium .11.
domus. Eodem modo de gradibus limbi inter eam in ascendente et
punctum in angulo terre facies, et habebis alias .3. domos, scilicet,

¹ Pt II. § 36.   ² Pt II. § 37.

inicium secunde [et] .3. domus. Nadayz autem istarum sunt inicia 396
sex oppositarum domorum.

De aspectibus planetarum.

Si autem aspectus duorum planetarum, uel .2. graduum quorum-
libet scire uolueris, pone eandem regulam super ipsos, et vide graduo 400
limbi intermedios, qui si fuerint .60., est aspectus sextilis; si .90.,
quartilis; si .120., trina; si .180., oppositionis; si nichil fuerit,
coniuncti. Si autem citra hos terminos .5. minus fuerit, erit appli-
catio ad aspectum; si plus, separatio ab eodem. Secundum quosdam, 404
hiidem aspectus habentur ex gradibus equalibus. Secundum ptho-
lomeum fit aliter, secundum gradus ascencionum, quemadmodum
equatio domorum sic atque numeris. Radiationum alia dextra, alia
sinistra; pro sinistra quidem radiatione, gradum planete super lineam 408
meridianam pone, atque almuri signa; deinde ipsum almuri motu
dextro, pro radiatione exagonali, .60. gradus procedat; pro tetragonali,
.90.; pro triagonali, .120.; et notetur medii celi gradus, ipse enim
radiationis prime locus est; deinde gradum planete super almucan- 412
therath orientale pone, atque almuri signa, procedatque almuri motu
dextro pro exagonali quidem .60., pro triagonali .120., pro tetragonali
.90.; et notetur gradus ascendens, ipse enim radiationis secunde
locus est; accipe itaque differentiam istarum duarum radiationum, et 416
serua eam. Deinde gradum medii celi hora acceptionis operis super
meridianum pone, et signetur alius; procedatque motu dextro, donec
planete gradus meridiano insideat, fiatque nota in almuri et capiatur
numerorum .2. intersticium, ducaturque in differentiam radiationum; 420
quodque inde producetur per arcum lucis siue diei ipsius planete
diuidatur, si super terram fuerit radiatio planete; si uero sub terra,
per arcum noctis eius; et quod de diuisione exierit, erit radiationis
equatio; que equatio minuetur a radiatione maiori, si fuerit planeta 424
inter .10. et .7. aut inter .4. et primum; addetur equatio super
radiationem minorem; sicque post additionem [vel] subtractionem
habebis radiationem quesitam; pro dextra autem radiatione inueni-
enda, erit processus almuri motu sinistro promouendus; cetera ut 428
supra.

Scientia anni mundani vel naturalis.

Cvm uolueris anni naturalis vel meridiani reuolutionem scire,
432 gradum ascendentis transacti anni pone super orizontem in oriente,
et locum almuri in margine signa ; posthec almuri ab eodem loco in
.93. gradu moue, et gradus qui ceciderit supra orizontem est gradus
ascendentis eiusdem anni.   Si autem planetes fuerint anni, pro vno-
436 quoque anno reduces almuri .93. gradus, et gradus existens in orizonte
in parte orientali erit ascendens ipsius anni.

Quot hore equalis sunt inter annum preteritum et reuolutum.

Si autem uolueris scire quot hore equales sint inter annum pre-
440 teritum et annum reuolutum, gradum perambulationis almuri diuide
per .15., et numerus qui exierit de diuisione est numerus equalium
horarum inter utrumque annum exientium.

De Gnomonis officio ; et primo, de vmbra altitudinis.

444   Qvadrantis in astrolabio constituti .2. sunt latera, in .12. partes
equales diuisa, que uocantur puncta umbre.   Sed notandum, quod
latus inferius uocatur umbra extensa ; et aliud latus vmbra uersa ;[1]
quia vnum representat puncta vmbre extense, et aliud uerse.   Cum
448 ergo per hoc opus uolueris scire quot punctorum gnomonis fit umbra
uersa vel extensa, considera altitudinem solis ; si fuerint .45. graduum
est vnaqueque earum .12. punctorum equalium, scilicet, suo gnomoni.
Si autem fuit máior altitudo solis, tanget regula latus vmbre extense ;
452 et si diuiseris per ea .144., inuenies puncta uerse.   Si uero solis altitudo fuerit minor .45. graduum, tactus regule in vmbra uersa ostendet
eius puncta ; per que diuide, et habebis puncta umbre extense ; nam
si puncta umbre uerse multiplicaueris in puncta umbre extense,
456 prouenient ex multiplicatione .144., que proueniunt eciam ex multiplicatione .12. in semet ipsis, que sunt partes gnomonis vnius. Scien-
dum est eciam quod si in acceptione umbre per altitudinem ceciderit
regula in parte alicuius puncti, et uolueris eam denominare a toto,
460 moue regulam ab inicio illius puncti in partem ipsam, et vide quot

[1] Cf. Part I. § 12.

gradus moueatur regula, qui erunt gradus illius partis. Deinde moue
regulam ab inicio illius partis in finem illius, et vide iterum quot
gradus moueatur regula, qui erunt gradus totius; tanta proportione
se habet pars puncti ad totum punctum. 464

Inuencio altitudinis rerum per puncta vmbre; capitulum.

Ut autem per umbram inuenias altitudinem, pone regulam supra
puncta vmbre extense, si fuerint pauciora .12., et tactus eius in
quarta altitudinis ostendet altitudinem. Si autem fuerint plura .12., 468
diuide per ea .144., et inuenies puncta umbre verse; super que pone
regulam, et tactus eius in quarta altitudinis ostendet tibi altitudinem.
Si fuerit vmbra .12. punctorum, est altitudo .45. Si uero cum pre-
dictis habueris fractiones, vide quid debeatur sibi de gradibus, ut 472
supra demonstratum est.

Inuencio vmbre meridiei per altitudinem.

Cum uolueris medie diei umbram scire, altitudinem solis in medio
eiusdem diei quere, et per ea inuenies umbram, ut supra dictum est. 476

Inuencio altitudinis rei accessibilis sequitur apponenda.

Cvm eleuate rei altitudinem uolueris scire, regula[m] super .45.
gradum in quarta altitudinis pone, et tam ante uel retro moue, donec
per vtriusque tabule foramen rei eleuate videas summitatem; tunc 480
quanta est longitudo a loco in quo fueris in radicem rei, cum additione
stature tue a visu usque ad terram, tanta est procul dubio altitudo
rei. [1]Si autem eius altitudinem, ita ut non remouearis a loco vno,
uolueris inuenire, tam diu regulam subleua uel deprime quod per 484
utriusque foramen videas cacumen; tunc si regula ceciderit super
puncta umbre extense, considera quanta proportione se habeant .12.
ad ista puncta; et tanta proportione se habebit altitudo rei ad longitudinem inter te et ipsam, cum statura tua addita longitudini. Si 488
uero ceciderit super puncta umbre uerse, quota pars erunt puncta de
.12., tanta pars erit altitudo rei illius longitudinis inter se et eius
radicem, coniuncta longitudini statura tua. Vnde notandum, quod

---

[1] Part II. § 41.

492 si fuerit regula super dyametrum quadrantis, est rei altitudo equalis longitudini, sibi addita statura. Et si fuerit super umbram extensam, est altitudo maior longitudine; si uero est super uersam, minor longitudine.

496   ### De altitudine rei inaccessibilis mecienda capitulum.

[1]Si uero rei inaccessibilis altitudo fuerit metienda, per utrumque regule foramen metiende rei summitatem respice, quia inspecta puncta quot sint mecientur, que, exempli causa, dicantur .3., que in 500 latere umbre quater continentur; quo pacto, retro ab eodem loco perge, ut mensurande rei cacumen iterum per utrumque foramen videas; quo viso, numerum punctorum umbre denuo vide, que scilicet erunt .2. puncta, que in .12. punctis continentur sexies; et 504 interuallum stationum .12. pedum notabis esse. Hijs itaque pactis, minus continens ternarij, scilicet .4., a maiori continente binarij, scilicet .6., auferatur, et binarius qui pertransierit memorie commendetur; et interuallum .2. stationum, quia ex proportionibus 508 remansit binarius, duplum altitudini inaccessibili pro certo habeatur. Est enim omnibus hec vniuersalis regula: subtractione continencium facta, si unum remanserit, interuallum stationum metientis erit altitudini rei equalis; si duo, duplum; si tria, triplum; et sic de 512 ceteris intellige.

De mensuracione plani: Rubrica.

Si queris cum astrolabio metiri planiciem, per utrumque foramen limitem eius ex aduerso posito considera; post hec puncta umbre supra 516 [quam] steterit regula ad .12. compara; et qualis fuerit comparatio punctorum ad .12., talis est comparatio stature tue ad planiciem.

Explicit astrolabium messehalle.

---

[1] Part II. § 42. As here expressed, it is suitable only for the *umbra versa*, because of the expression "retro perge"; but it may easily be altered so as to suit the *umbra recta*, which would give § **43** in Chaucer's treatise. Such an additional section does, in fact, appear in MS. Hh. 6. 48, in the margin.

NOTES.

OBSERVE that I have only printed here the latter part of the treatise; see the Preface, p. xxiv. The former part is longer and less to the point.

I add a few notes on some of the readings of the MSS.; denoting MS. Ii. 3. 3 by the letter S, and MS. Hh. 6. 8 by T, to avoid confusion with the MSS. already mentioned.

L. 3. alhabor; *so in* T. But written *alhaboz*, very plainly, in S; indeed, the scribe clearly confused *r* with *z*, for he writes *nadayz* throughout for *nadayr*, or (as in l. 54) he writes *nadays;* cf. footnote to l. 261.

14. inter T; in S. So also in l. 276.

37. nadayz S; nadayr T; see note to l. 3.

126. *allidadam;* i. e. the "rewle" for taking altitudes. See footnote on p. 7; and the first Additional Note to Part ii, sect. 3.

155. *Quis dies (sic).* But in the next line it is *que dies.*

247. inter uxorem minorem latitudinem S; but T omits *uxorem.* The scribe seems to have been thinking of something else besides his work.

299. Perhaps we should read *fractionibus,* or *cum fractionibus.* The passage is wanting in T.

329. opp*a*ruit (*sic*) S; apparu*e*rit T.

342. poteris S; poterit T.

461. gradus moueatur; *for* gradus, T *has* gradibus.

500. retro T; recto S.

506. auferatur T; auferantur S.

516. quam; *supplied from* T.

GLOSSARIAL INDEX.

ABATE, 2 *p. s. pr. subj.* subtract, ii. 10. 8.

Abid, *imp. s.* wait, ii. 23. 9.

Abowe, *prep.* above, ii. 45. 32.

Acording, *pres. part.* agreeing, ii. 14. 5.

Addyng, *sb.* (the) addition, ii. 41. 9.

Aftur, *prep.* according to, *pr.* 57 ; later than, i. 21. 15 ; After, *pr.* 62.

Agayn-ward, *adv.* backward, at the point of return, i. 17. 8.

Agayns, *prep.* against, near to, ii. 23. 8.

Al be it so þat, although, ii. 31. 6.

Almenak, *sb.* almanac, *pr.* 62. The real origin of this word is still unknown; it is probably *not* of Arabic origin, and the fact of its beginning with *al* has probably led inquirers astray. The word occurs in a passage in a lost work of Porphyry, cited by Eusebius, de Præpar. Evangelica, iii. 4. ed. Gaisford. See Dozy, Glossaire des mots Espagnols dérivés de l'Arabe ; 2nd ed. p. 154.

Almicanteras, *sb. pl.* small circles of declination (in the celestial sphere), i. 18. 2 ; Almykan-teras, i. 18. 7. Arabic *muqantarat*, a solar quadrant, solar clock; pl. *muqantarát*, circles parallel to the horizon; from *qantara*, he bent.

Almury, *sb.* the "denticle" or tooth-like point or pointer situate on the Rete near the "head" of Capricorn, i. 23. 1. Arabic *al-murî*, the shewer, part. of the 4th or causal conjugation of *raá*, to see.

Altitude, *sb.* the elevation of a celestial object above the horizon, measured along a vertical arc, *pr.* 56.

Alyne, *adv.* in an exact line, ii. 38. 16.

Amenuseth, *pr. s.* diminishes, becomes less, i. 21. 45. O.F. *amenuiser*, to become less, from *menut, menu*, minute, small.

A-middes, in the midst of, i. 18. 3.

Amydde, *adv.* amidst, in the middle, i. 4. 4.

Angle, *sb.* angular distance (from the meridian), ii. 4. 28. "*Angle* of longitude (in astrology) is the angle which the circle of a star's longitude makes with the meridian, at the pole of the ecliptic."— Bailey. This is not explicit, as the pole of the ecliptic is seldom in the meridian; the pole of the equator and a circle of right ascension would answer better.

Anni collecti, collected years, ii. 44. 16. When a table contains quantities denoting the amount of a planet's motion during round periods of years, such as 20, 40, or 60 years, such a change is entered under the heading *Anni Collecti*.

Anni expansi, expanse years, ii. 44. 16. When a table contains quantities denoting the amount of a planet's motion during only a few years, viz. from 1 to 20 years, such changes are entered separately under the headings 1, 2, 3, &c. years, which are designated the expanse (or separate) years.

Antartik, *adj.* southern, ii. 25. 7.

Anything, *adv.* in any degree, at all, to any appreciable extent, ii. 17. 6; Anythyng, ii. 38. 9.

Aperceyue, *v.* to perceive, ii. 35. 4.

Arabyens *sb. pl.* Arabians; *hence,* in Arabyens, among the Arabians, i. 10. 5.

Arch, ii. 9. 2. *See* Ark.

Areisid, *pp.* raised, ii. 2. 5.

Ariste, *sb.* arising, rising, ii. 12. 10.

Ark, *sb.* arc, the angular distance apparently passed over by the sun during a day or a night, ii. 7. 7, 8; Arch, ii. 9. 2; *pl.* Arches, ii. 7. 9.

Armholes, *sb. pl.* i. 21. 51.

Artificial, *adj.* ii. 7. *rub.* The *day artificial* is the length of the day, from the moment of sunrise to that of sunset.

Artik, *adj.* arctic, northern, ii. 22. 2.

Ascendit, *pt. s.* ascended, rose above the horizon, ii. 40. 49; Assendid, ii. 40. 28.

Ascensioun, *sb.* ascension, ii. 26. 4; *pl.* Ascensiouns, ii. 26. 2. For the signs of *right* and *oblique ascension* see note on pp. 35, 36.

Aspectys, *sb. pl.* aspects, ii. 4. 30. An *aspect* is the angular distance between two planets. The principal aspects are five, viz. conjunction, sextile, quartile, trine, and opposition, corresponding to the angular distances 0°, 60°, 90°, 120°, and 180° respectively.

Assendent, *sb.* ascendent, ii. 3. 24. The "ascendent" is that point of the ecliptic which, at a given moment, is ascending above the horizon.

Astrelabie, *sb.* an astrolabe, *pr.* 4; Astralabie, *pr.* 7; Astrolabie, *pr.* 46.

Astrologie, *sb.* astrology, *pr.* 70.

Astrologien, *sb.* astrologer, *or rather,* astronomer, *pr.* 50; *pl.* Astrologiens, *pr.* 42.

Atones, *adv.* at once, at one and the same time, *pr.* 32.

Auctours, *sb. pl.* authors, ii. 39. 22.

Auenture, *sb.* chance; per auenture, peradventure, perchance, perhaps, ii. 12. 6.

Augrym, *sb.* arithmetic, numeration, i. 7. 4; Augrim, Arabic numerals, i. 8. 4. O.F. *algorisme,* O.Sp. *alguarismo,* either from Gk. ἀριθμός, number, with the Arabic article (*al*) prefixed; or, as is much more probable, from *Al Khowáresmí,* the surname of an Arabian writer on algebra, the translation of whose work was the means of introducing the decimal notation into Europe in the 12th century. See the authorities for this in Dozy, Glossaire des mots Espagnols dérivés de l'Arabe.

Avisely, *adv.* advisedly, carefully, ii. 29. 16.

Awaite, *imp. s.* watch, ii. 35. 6.

Azymuthz, *sb. pl.* azimuths, i. 19. 4, &c. Arabic *as-samt,* pl. *as-sumút,* a way or path; *al-samt,* a point of the horizon, and hence, an *azimuth,* or arc extending from the zenith to the horizon. We find also *samt al-ras,* the vertex of heaven; hence *zenith,* a corruption of *semt* (*al-ras* being dropped).

Bakhalf, *sb.* the back or flat side of the Astrolabe, i. 4. 1; ii. 1. 6.

Bak-side, *sb.* the back of the Astrolabe, i. 15. 3.

Bas, *sb.* base, ii. 41. 2 ; Baas, ii. 43. 2.

Be, *prep.* by, ii. 41. 3.

Befornseyd, *pp.* aforesaid, ii. 42*b*. 5.

Behete, 1 *p. s. pr.* promise, *pr.* 73. A.S. *behát*, a promise, *be-hátan*, to promise.

Bestes, *sb. pl.* the "beasts" or animals in the zodiacal signs, i. 21. 37.

Bisily, *adv.* diligently, ii. 38. 8.

Bordure, *sb.* outer border or raised rim on the front side of the astrolabe, i. 4. 2 ; 16. 1 ; ii. 38. 2.

Brede, *sb.* breadth, i. 21. 24.

Brodere, *sb.* larger, ii. 38. 1.

By, *prep.* with reference to, ii. 3. 48.

Byforn, *prep.* before, ii. 3. 15.

Byhestes, *sb. pl.* promises, i. e. what they profess to prove, *pr.* 16. A.S. *behǽs*, a promise.

By-twyxe, *prep.* between, ii. 28. 11, 14.

Caas, *sb.* case ; sette caas, suppose, ii. 42. 14.

Calkuled, *pp.* calculated, *pr.* 52.

Calle, *sb.* caul, a net used to confine women's hair, i. 19. 3. "Maydens were sylken *callis*, with the whiche they kepe in ordre theyr heare, made yelowe with lye ;" Hormani Vulgaria, leaf 115.

Canon, *sb.* a rule, explanation, *pr.* 63 ; Canoun, a canon, rule, table, ii. 32. 3.

Capitalles, *adj. pl.* capital, ii. 3. 20 ; Capitals, i. 16. 8.

Centre, *sb.* the "centre" or small point at the very end of which is the position of a fixed star in the Rete of the Astrolabe, i. 21. 7.

Cenyth, *sb.* (1) zenith, i. 18. 10 ;

(2) a word denoting the point where a given azimuth-circle meets the horizon, i. 19. 7. *See* Azymuthz, *and* Scnyth.

Certein, *adj.* certain ; a certein (*used without a sb.*), *pr.* 10 ; a certein holes, i. e. a certain number of holes, i. 13. 2. So also 'of unces a certain,' C. T. 16244 ; 'a certain of gold,' C. T. 16492.

Chapitre, *sb.* chapter, ii. 9. 2.

Clepen, *pr. pl.* name, mention, ii. 39. 23 ; *pp.* Cleped, named, i. 4. 4.

Clokke, *sb.* clock, *pr.* 56.

Clymat, *sb.* a belt or zone of the earth included between two given lines of latitude, ii. 39. 18 ; Climat, ii. 39. 19 ; *pl.* Clymatz, climates, i. e. zones of latitude, i. 3. 3 ; Clymates, sets of almicanteras calculated for various terrestrial latitudes, i. 14. 2.

Combust, *adj.* quenched, viz. by being so near the sun as to be obscured by his superior light, ii. 4. 31.

Compas, *sb.* broad circle, zone, i. 21. 23 ; a circle, ii. 38. 2.

Compassed, *pp.* drawn with compasses, fashioned circularly, i. 18. 1.

Compilatour, *sb.* compiler, *pr.* 42.

Compowned, *pp.* compounded, i. e. composed, constructed, drawn, *pr.* 7 ; described, marked, i. 18. 8 ; Compownet, ii. 5. 2.

Comunly, *adv.* commonly, ii. 19. 7.

Conclusions, *sb. pl.* mathematical propositions, either problems or theorems, *pr.* 9.

Conforted, *pp.* comforted, supported, strengthened, ii. 4. 29.

Coniunccion, *sb.* conjunction, ii. 32. 1. It means a very close apparent approach of two celestial bodies.

Consentrik, *adj.* having the same

centre, i. 17. 3; Consentryk, i. 17. 34; Consentrik, tending to the same centre, i. 16. 5; at an unchanging altitude, ii. 3. 47.

Contienen, v. to contain, pr. 53; pr. s. Contienith, i. 7. 7; pr. pl. Contienen, i. 9. 2.

Cost, sb. quarter, direction, ii. 46. 5; Coste, ii. 46. 17; pl. Costes, coasts, directions, i. e. parts of the sky to be observed, i. 19. 6.

Cowchyng, sb. laying down, letting the Astrolabe lie flat on the ground, ii. 29. 18.

Crepusculus, sb. pl. twilights, durations of twilight, ii. 6. rub.; Crepusculis, ii. 9. 1.

Croos-lyne, sb. cross-line, the line from right to left through the centre in Fig. 1; i. 12. 5.

Curious, adj. ornate, pr. 31.

Dawenyng, sb. dawning, dawn, ii. 23. 8.

Dawyng, sb. dawning, ii. 23. 20.

Declaracioun, sb. explanation, i. 3. 3, &c.

Declinacioun, sb. the angular distance of a celestial obiect N. or S. of the equator, measured along an arc at right angles to it, i. 17. 4; pl. Declinacions, pr. 54.

Declinen, pr. pl. possess declination, i. e. pass either to the north or south of the ecliptic, ii. 17. 28; pr. s. Declineth, possesses declination, ii. 19. 8.

Denticle, sb. pointer, i. 23. 1. See Almury.

Departen, pr. pl. separate, i. 21. 32; imp. s. Departe, divide, ii. 8. 2; pr. s. Departeth, divides, i. 17. 30; pp. Departid, divided, marked by lines, i. 21. 34; divided, ii. 4. 37.

Depressioun, sb. the angular distance of the southern pole from the horizon, ii. 25. 6.

Descencioun, sb. descension, ii. 4. 32; Discencioun, ii. 4. 33. The technical signification seems to be —the "house" or portion of the sky just above the western horizon, so that perhaps a planet in his descension is about to set. (?)

Descriued, pp. described, marked, i. 17. 1.

Desturbith, pr. s. disturbs, prevents, i. 2. 2.

Determynat, adj. properly ascertained, i. 21. 4; properly placed upon the Astrolabe, ii. 18. rub.

Dignite, sb. dignity, ii. 4. 29; pl. Dignetes, pr. 72. A term in astrology. Bailey defines Dignities as "the advantages which a planet has on account of being in a particular place in the zodiac, or in such a station with other planets, &c."

Directe, adj. direct, ii. 35. 11. A planet's motion is direct when it moves in the same direction as the sun along the signs.

Directe, in directe, in a line with, ii. 44. 9.

Distantz, adj. pl. distant; euene distantz, equidistant, i. 17. 31.

Doctrine, sb. instruction, pr. 43.

Downere, adv. more downward, ii. 12. 14.

Ebrew, adj. Hebrew, pr. 23.

Ecliptik, sb. ecliptic, pr. 67. A great circle of the sphere, drawn along the middle of the zodiac, making an angle with the equator of about 23°. 28'; the apparent path of the sun, nearly.

Eft-sones, adv. soon after, immediately, ii. 23. 11.

Egge, sb. edge, ii. 46. 7.

Elecciouns, sb. pl. elections, choice of fit times, ii. 4. 2. "Of viage is ther non eleccioun;" Chaucer; M. of L. Ta. 312.

Eleuacioun, *sb.* the altitude above the horizon of the north pole, ii. 23. 15.

Eleuat, *pp.* elevated, ii. 23. 17. *See* Eleuacioun.

Elles, *adv.* otherwise, i. 19. 2.

Elongacioun, *sb.* angular distance, ii. 25. 39.

Embelif, *adj.* oblique, i. 20. 2 ; (as applied to angles), angles that are not right angles, ii. 26. 23. I cannot find the word elsewhere, nor can I guess to what language it belongs; it looks like an extraordinary corruption of the word *oblique.*

Embelif, *adv.* obliquely, ii. 26. 6.

Emysperies, *sb. pl.* hemispheres, i. 18. 6.

Endlang, *adv.* along, lengthways, ii. 40. 22, 44.

Endytyng, *sb.* inditing, style of composition, *pr.* 30.

Engin, *sb.* ingenuity, skill, *pr.* 41. Lat. *ingenium,* whence F. *engin.*

Enhausyng, *sb.* elevation, ii. 39. 16.

Enhawsed, *pp.* exalted, elevated, lifted above (the horizon), ii. 26. 22. O.F. *enhaucier,* to elevate, *haut,* high.

Ensample, *sb.* token ; in ensample, to signify, i. 21. 25.

Ensampulle, *sb.* example, ii. 45. 5.

Entere, *imp. s.* enter, ii. 44. 7. To "enter with" is to keep in mind and search for, as a help to finding something else. "*Argument,* in astronomical tables, is the angle on which the tabulated quantity depends, and with which, therefore, in technical language, the table must be *entered.*"—Eng. Cycl. Arts and Sciences, s. v. *Argument.* In l. 3. *entere hit* = set down in writing.

Entres, *sb. pl.* entries, ii. 44. 26.

Episicle, *sb.* epicycle, ii. 35. 18.

A small circle, the centre of which moves along the circumference of a larger one.

Equacion, *sb.* equal partition, ii. 37. 9 ; *pl.* Equaciouns, ii. 36. 2. The "equations of houses" means the method of dividing the sphere *equally* into "houses" for astrological purposes; Equacions, *pr.* 71 ; Equaciouns, calculations, i. 23. 3.

Equales, *adj. pl.* of equal length ; howris equales, hours each containing 60 minutes, ii. 8. 2 ; Equals, equal, i. 16. 9.

Equinoxial, *adj.* equinoctial ; said of the equinoctial circle or equator, the great circle of the sphere whose poles are the arctic and antarctic poles, i. 17. 12.

Equinoxiis, *sb. pl.* equinoxes, i. 17. 18.

Est, *adj.* East, i. 5. 4.

Eue, *sb.* evening, ii. 12. 22.

Euene, *adv.* equally. *See* Distantz.

Euerech, *pron.* every one, *pr.* 40.

Euer-mo, evermore, ii. 3. 25.

Euidently, *adv.* by observation, ii. 23. *rub.*

Expanse, *adj.* expanse or separate, ii. 45. 11. *See* Anni expansi.

Experience, *sb.* knowledge acquired by trial, ii. 1. 16.

Extre, *sb.* axle-tree, i. 14. 1. A.S. *eax,* an axis, axle.

Face, *sb.* a third part of a "sign," a portion of the zodiac 10 degrees long, ii. 4. 38.

Failling, *pres. part.* failing, remote, ii. 4. 18.

Farwel ; go farwel, be dismissed, be let alone, ii. 23. 7.

Felicite, *sb.* favourable position or aspect, ii. 4. 25.

Fer, *adj.* far, ii. 16. 1.

Ferforth, *adv.* far-forth, i. e. far, *pr.* 49.

Ferthe, fourth, ii. 35. 4.

Ferther-ouer, *conj.* moreover, ii. 26. 8.

Figures, *sb. pl.* figures, i. e. markings, *pr.* 45.

Fixe, *pp.* fixed, *pr.* 54.

Fond, 1 *p. s. pt.* found, ii. 1. 6.

For, *prep.* against, to prevent, ii. 38. 1; to have for excused, i. e. to excuse, *pr.* 30.

Forþer, *adv.* further, ii. 43*a*. 4.

Forthward, *adv.* forwards, ii. 35. 5.

For-why, *conj.* because, ii. 46. 19.

Frere, *sb.* friar, *pr.* 58.

Fro, *prep.* from; fro vs-ward, away from us, used to express that the sun having reached the nearest point to our zenith, begins to descend from it, i. 17. 9. Cf. i. 17. 39.

Furth, *adv.* forward, ii. 46. 4; Furþe, ii. 46. 16.

Geuen, *pp.* given, *pr.* 6.

God, *adj.* good, ii. 4. 28.

Gouernance, *sb.* regulation, *pr.* 56; subjection, i. 21. 50.

Grek, *adj.* Greek, *pr.* 22; *pl.* Grekes, *pr.* 20.

Gyrdelle, *sb.* girdle, cincture, central line or great circle, i. 17. 26; Girdel, i. 17. 29.

Haddy, *for* Hadde y, had I, ii. 1. 15.

Halidayes, *sb. pl.* holydays, i. 11. 1.

Halt, *pr. s.* holdeth, holds, i. 14. 2.

Han, *pr. pl.* have, possess, *pr.* 24.

Hastow, *for* hast thou, i. 5. 6.

Hath hymself, *pr. s.* bears a ratio. is in proportion, ii. 41*b*. 5.

Hauy, *for* haue y, have I, ii. 40. 15.

Hedes, *sb. pl.* heads, or first points of signs, i. 17. 12. *See* Heued.

Heie, *adj.* high, i. 16. 7.

Heiest, *adj.* highest, ii. 13. 4.

Hem, *pron. pl.* them, i. 8. 7.

Hennes-forthward, *adv.* henceforth, i. 1. 3.

Hepe, *sb.* heap; *hence,* to hepe, in a heap, all close together, i. 14. 4. See the Preface, p. xxviii.

Her-mele, *sb.* the thickness of a hair, a hair's breadth; lit. a hairpart, ii. 38. 10. A.S. *mæl,* a portion.

Heued, *sb.* head; the beginning or first point of a zodiacal sign, i. 17. 3; *pl.* Heuedes, i. 17. 16.

Heuenissh, *adj.* heavenly, i. 21. 35. Cf. Compl. of Mars, st. 5.

Heuy, *adj.* heavy, difficult, *pr.* 32.

Heyer, *adj.* higher, ii. 23. 26.

Heyhte, *sb.* height, altitude, i. 1. 2; ii. 3. 13; Heyȝte, ii. 41. 9; Heyȝth, ii. 41. 13.

Hihten, *pr. pl.* are called, i. 18. 2.

Hir, *pron.* her (applied to a star), ii. 3. 29.

Hir, *pron.* their, *pr.* 16; i. 21. 4.

His, *pron.* its, i. 2. 3.

Hise, *pron. pl.* his, i. 12. 3.

Hit, *pron.* it, i. 2. 2.

Hole, *adj.* whole, ii. 9. 3.

Horoscopo; in *horoscopo,* within that part of the sky considered as the ascendent, ii. 4. 8; see note on p. 18. Gk. ὡροσκόπος, observing hours; also, as sb., a nativity, a horoscope; from ὥρα, time, an hour, σκέπτομαι, to consider.

Horoscopum, *sb.* horoscope, ii. 4. 36. *See* above.

Hors, *sb.* the "horse," a name for the little wedge that passes through a hole in the end of the "pyn," i. 14. 4. Called in Arabic *alpheraz*, the horse.

Howis, *sb.* house, ii. 36. 5; Howys, ii 36. 7; *pl.* Howses, *pr.* 71. The whole celestial sphere was divided into twelve equal portions, called *houses*, by six great circles passing through the north and south points of the horizon; two of these circles being the meridian and the horizon.

Ilike, *adj.* equal, i. 17. 31; Illike, like, equal, i. 17. 17. A.S. *gelíc*, G. *gleich*.

Illik, *adv.* equally, ii. 15. 1; the same, ii. 39. 13.

Illike-distant, at an even distance, i. e. parallel, ii. 39. 18. (*Not* equidistant, because the climates varied in breadth.)

In, *prep.* into, i. 16. 2; among, i. 10. 5.

Indeterminat, *adj.* not marked upon the Astrolabe, ii. 17. *rub.*

Inequal, *adj.* ii. 10. 4; *pl.* Inequales, of unequal length; howris inequales, hours formed by dividing the duration of daylight by twelve, ii. 8. 1; Inequalis, ii. 10. 1.

Infortunat, *adj.* unlucky, ii. 4. 34.

Infortunyng, *sb.* unlucky condition, ii. 4. 26.

Inperfit, *adj.* imperfect, incomplete, i. 18. 3.

Intercept, *pp.* intercepted, ii. 39. 24.

Introductorie, *sb.* introduction, *pr.* 68.

Ioigned, *pp.* joined, nearly or altogether in conjunction, ii. 4. 31.

Ioyntly, *adv.* conjointly, together, ii. 11. 9.

Iudicial, *adj.* judicial, ii. 4. 35. *Judicial astrology* pretended to forecast the destinies of men and nations; *natural astrology* foretold natural events, such as the weather and seasons.

Iust, *adj.* just, exact, ii. 3. 43.

Iustly, *adv.* exactly, ii. 3. 44.

I-wreten, *pp.* written, ii. 45. 22: I-wrete, ii. 45. 23.

I-wryton, *probably an error for* I wolde witen, I would know, ii. 45. 6. See the Critical Note.

Kalcule, *v.* to calculate, i. 22. 3.

Kalender, *sb.* a calendar, i. 11. 1; *pl.* Kalendres, *pr.* 57. Lat *calendarium.* The old calendars answered nearly to our modern almanacks.

Kalkuler, *sb.* the calculator or pointer, i. 23. 2. *See* Almury.

Kanstow, 2 *p. s. pr.* knowest thou, *pr.* 20.

Kas, *sb.* case; in kas þat, in case, ii. 3. 2.

Kawht, *pp.* caught, perceived, ii. 17. 8.

Kep, *sb.* heed; tak kep, take heed, i. 1. 2.

Keruyng, carving, i. e. cutting, crossing over, i. 19. 3.

Knowyng, *sb.* knowledge, *pr.* 47.

Kon, *imp. s.* grant; kon me thank, grant me thanks, thank me, *pr.* 38. "To con one thanks, Fr. *savoir gré,* to feel thankful and make the feeling known to the object of it."—Wedgwood. A.S. *cunnan*, to know.

Kowch, *v.* to lie; kowch adown, lie down, ii. 29. 14.

Krokede, *adj.* crooked, i. 19. 2.

Label, *sb.* the narrow revolving rod or rule on the front of the Astrolabe, i. 22. 1. *See* Fig. 6.

Lasse, *adj.* less, *pr.* 40.

Lat, *imp. s.* let, ii. 29. 13.

Latitude, *sb.* breadth (*without any astronomical sense*), i. 21. 26; the breadth of a "climate;" or rather, a line along which this breadth is measured, ii. 39. 19. *See* below.

Latitude, *sb.* (1) *astronomical;* the angular distance of any body from the ecliptic, measured along a great circle at right angles to the ecliptic, *pr.* 66; (2) *terrestrial*, the distance of any place on the globe, N. or S. of the equator, ii. 39. 23; (3) the breadth of a "climate," ii. 39. 19.

Leden, *pr. pl.* lead, conduct, *pr.* 28.

Lengere, *adj. pl.* longer, ii. 10. 2.

Lengthing, *pres. part.* extending, ii. 25. 39.

Leoun, *sb.* Leo (the sign), ii. 25. 27. From Lat. acc. *leonem.*

Lest, *impers. pr. s.* it pleases, ii. 25. 38. *See* List.

Leste, *adj.* least, i. 17. 2.

Leuyth, *pr. s.* remains, ii. 25. 14; Leueth, ii. 25. 16.

Lewd, *adj.* unlearned, *pr.* 42.

Lewyth, *pr. s.* remains, ii. 44. 28. *See* Leuyth.

Ligge, *pr. s. subj.* may lie, ii. 41. 3.

Lihte, *adj. pl.* light, i. e. easy, *pr.* 19; *dat. sing.* Lihte, *pr.* 35.

Liked, *pt. s. impers.* it pleased, i. 10. 6.

List, *pr. s. impers.* it pleases (thee), ii. 3. 1. *See* Lest.

Lite, *adj. as sb.* a little, ii. 1. 15. A.S. *lyt.*

Lite, *adj.* little, *pr.* 20.

Lite, *adv.* a little, ii. 12. 8.

Longitude, *sb.* the distance between two given meridians, ii. 39. 12; the length or extent of a "climate," in a direction parallel to the equator, or rather (as it would appear), a line along which to measure this length; ii. 39. 18.

Longitudes, *sb. pl.* longitudes, *pr.* 53, 55. The longitude of a star is measured along the ecliptic; that of a town, from a fixed meridian.

Loppe, *sb.* a spider, i. 3. 4; 19. 2. A.S. *lobbe*, a spider.

Lop-webbe, *sb.* cobweb, i. 21. 2. *See* Loppe.

Lyhtly, *adv.* easily, ii. 14. 8.

Lyne, *sb.* a line, cord, ii. 23. 25.

Lyne-riht, *adj.* in an exact line, exactly in a line with, i. 21. 18.

Maistow, *pr. s.* mayest thou, i. 21. 46.

Maner, *sb.* kind; *used without* of following, *as* maner turet, kind of "turet," i. 2. 1; maner strikes, sort of strokes, i. 19. 1.

Matiere, *sb.* matter, subject, ii. 4. 35.

Mechel, *adv.* much; for as mechel, for as much, *pr.* 4.

Mediacion, *sb.* means, assistance, *pr.* 8; Mediacioun, use, i. 13. 3.

Membres, *sb. pl.* parts, *pr.* 46.

Mene, *adj.* mean, ii. 44. 13. *See* Mote.

Meridian, *adj.* meridional, at the moment of southing, exact southern, *pr.* 56; southern, on the meridian, ii. 39. 6.

Meridional, *adj.* southern, i. 4. 4.

Mete, 1 *p. s. pr.* measure, ii. 41. 5.

Michel, *adv.* much, ii. 23. 17.

Mile-wey, *sb.* a space of 5 degrees, which answers to 20 minutes of time, the average time for walking a mile; hence the term, i. 7. 7; *pl.* Mile-wey, i. 16. 10.

Minutes, *sb. pl.* (1) minutes of time, i. 7. 8; (2) Minute, i. e. a sixtieth part of a degree, i. 8. 8; see i. 8. 10.

Mo, *adj.* more, *pr.* 26.

Moder, *sb.* lit. mother; the thickest plate forming the body or principal part of the Astrolabe; called in Latin *mater* or *rotula*, i. 3. 1.

Modur, *sb.* mother, *pr.* 73.

Moeble, *adj.* movable, i. 21. 47.

Moeuyng, *sb.* moving; *pr.* 61; Moeuynge, *pr.* 59; firste Moeuyng, the "*primum mobile*," i. 17. 26.

Mone, *sb.* moon, *pr.* 61. A.S. *móna.*

Moneth, *sb.* month, ii. 44. 34; Monith, i. 10. 12; *pl.* Monythis, ii. 44. 33.

More, *adj.* greater, *pr.* 40; ii. 26. 7.

Morwe, *sb.* morning, ii. 12. 25.

Mote, *sb.* motion (Lat. *motus*), ii. 44. 13. The "mene mote" or *mean motion* is the motion of a planet during a given period as stated in the tables.

Nadir, *sb.* the point of the ecliptic exactly opposite to that in which the sun is situate, ii. 6. 1; see l. 8. Arabic *nadhíru's-samt*, i. e. opposite to the zenith, for which the term *an-nadhír* simply, signifying "opposite," was commonly used.

Naked, *adj.* simple, plain, *pr.* 19.

Nam, *for* Ne am, am not, *pr.* 42.

Narwe, *adv.* closely, lit. narrowly, *pr.* 49.

Narwest, *superl. adj.* narrowest, smallest, i. 18. 4.

Nat, *adv.* not, *pr.* 16.

Natheles, *conj.* not the less, never the less, *pr.* 20. A.S. *ná*, not.

Nativitez, *sb. pl.* nativities, castings of nativity in astrology, ii. 4. 1.

Nawht, *adv.* not, *pr.* 36.

Neer, *adv.* nearer, ii. 43*a*. 4; 42*b* 3; Ner, ii. 42. 3.

Nether, *adj.* lower, i. 12. 6.

Netherest, *adj. superl.* lowest, i. e.

outermost, i. 18. 4; Netherestc, lowest, i. 4. 2.

Neuer-mo, *adv.* never oftener, never (with two exceptions), ii. 31. 3.

Ney, *adj.* nigh, ii. 3. 46.

Nombre, *sb.* a number, *pr.* 9; amount, sum, ii. 24. 3; *pl.* Noumbres, *pr.* 2.

Notable, *adj.* noteworthy, *pr.* 57.

Noteful, *adj.* useful, *pr.* 72. A.S. *notu*, use.

Nowmbres, *sb. pl.* numbers, i. 7. 4. *See* Nombre.

O, one, one single, ii. 19. 11.

Obedient, *adj.* answering to, *or* subject to, ii. 28. 20. A technical term, applied to the eastern signs of the zodiac, as being respectively correspondent to the western ones.

Obeieth, *pr. s.* obeys, ii. 28. 25. *See* Obedient.

Occidentale, *adj.* Western, i. 5. 6. From Lat. *occidens*, setting.

Of, *prep.* by, *pr.* 41; for, i. 12. 4; from, i. 17. 28.

On, one, i. 10. 15; one o'clock, ii. 3. 50; in on, in one and the same condition, unchangeably, ii. 2. 8.

Ones, *adv.* once, *pr.* 34.

Onythyng, ii. 38. 12. *See* Anything.

Or, *prep.* ere, before, ii. 23. 20.

Orientale, *adj.* eastern, i. 5. 4. From Lat. *oriens*, rising.

Orizon rectum, or right horizon, ii. 26. 20. This means the horizon of any place situate on the equator, which could be represented by a *straight* line upon a disc or "table" of the Astrolabe.

Orizonte, *sb.* horizon, *pr.* 7. Lat. acc. *horizontem*; Gk. ὁρίζων, bounding.

Ouerkerueth, *pr. s.* cuts across,

crosses, i. 21. 53 ; Ouerkeruyth, ii. 26. 20.

Ouer-thwart, *prep.* exactly across, at right angles to, i. 5. 1. A.S. *þweorh,* across, diagonal. Cf. Knightes Ta. 1133.

Oxenford, *sb.* Oxford, *pr.* 8, 71.

Paiens, *sb. pl.* pagans, ii. 4. 35.

Parcelle, *sb.* parcel, i. e. part, i. 21. 49.

Partie, *sb.* part, *pr.* 45 ; *pl.* Parties, *pr.* 18.

Passeþ, *pr. s.* exceeds, ii. 42. 15 ; 42*a.* 7.

Perced, *pp.* pierced, i. 3. 2.

Perche, *sb.* a rod placed high up in a horizontal position, ii. 23. 26. Lat. *pertica.*

Perfit, *adj.* perfect, complete, i. 18. 2.

Perfitly, *adv.* perfectly, *pr.* 13.

Performe, *v.* to shew, constitute, be equivalent to, ii. 10. 10.

Peyre, *sb.* a "pair," a set, ii. 40. 18. A *pair* by no means implies that the set of similar things to which it is applied is limited to *two.* Cf. Prol. to Cant. Tales, 159.

Plages, *sb. pl.* quarters of the compass, i. 5. 7 ; ii. 31. 10. Lat. *plaga,* a region, space.

Planetes, *sb. pl.* planets, *pr.* 72. The seven planets, in order, are the Moon, Mercury, Venus, the Sun, Mars, Jupiter, and Saturn.

Plate, *sb.* the "sight" on the "rewle," i. 13. 2.

Pleie, *v.* to play ; *hence,* to use, apply, ii. 40. 57. A.S. *plegan,* to play, apply.

Plomet ; *sb.* plummet, heavy weight, ii. 23. 25.

Plom-rewle, *sb.* plummet-rule, ii. 38. 6.

Pol, *sb.* pole, i. 14. 6 ; Pool, i. 18. 12.

Portatif, *adj.* portable, *pr.* 50.

Practik, *sb.* practice, practical working, *pr.* 49.

Precedent, *adj.* preceding, ii. 32. 3.

Preue, *sb.* test, verification, experimental proof, ii. 23. *rub.*

Prikke, *sb.* a small mark, such as a little stick stuck in the ground, ii. 42. 3 ; a dot, ii. 5. 12.

Principals, *adj. pl.* principal, chief, i. 4. 7 ; Principalx, cardinal, ii. 31. 10.

Proporciouns, *sb. pl.* proportions, ratios, *pr.* 3.

Propre, *sb.* own, ii. 7. 14.

Propretes, *sb. pl.* properties, i. 10. 5.

Proue, *v.* to test, ii. 23. *rub.* Lat. *probare,* to test, verify.

Puttyng to, i. e. adding, ii. 43*a.* 12.

Pyn, *sb.* the pin which passes through the central hole in the Astrolabe and its plates, i. 14. 1.

Quantite, *sb.* largeness, size, i. 21. 24.

Rather, *adv.* sooner, i. 21. 14. A.S. *hræð,* quick.

Reherse, *v.* to rehearse, enumerate, *pr.* 45.

Remenant, *sb.* remnant, rest, i. 4. 5.

Remeue, *imp. s.* move backwards and forwards, ii. 2. 2.

Rennyth, *pr. s.* runs, continues, ii. 3. 47.

Resceiued, *pp.* received ; wel resceiued, favourably situated with respect to other planets, &c. ; ii. 4. 30.

Respecte, *sb.* regard, i. 21. 49.

Resseyuyth, *pr. s.* receives, i. 3. 2.

Retrograd, *adj.* moving in a

direction contrary to that of the sun's motion in the ecliptic, ii. 4. 31; 35. 12. Spoken with reference to a planet's *apparent* motion.

Reuerent, *adj.* reverend, *pr.* 58.

Revolucioun, *sb.* complete circuit, ii. 7. 13.

Rewde, *adj.* rude, plain, un-adorned, *pr.* 30.

Rewle, *sb.* the revolving long and narrow plate or rod used for measuring and taking altitudes, i. 13. 1. See Fig. 3. It revolves at the *back* of the Astrolabe.

Rewles, *sb. pl.* rules, *pr.* 19.

Riet, *sb.* the "rete" or net; the topmost plate on which some of the stars were figured, and the northern half of the zodiac shewn, i. 3. 3.

Rikened, 1 *p. s. pt.* reckoned, counted, ii. 3. 35.

Rond, *adj.* round, circular, ii. 38. 1; ronde, ii. 38. 2.

Rote, *sb.* root, the tabulated quantity belonging to a given fixed date, from which corresponding quantities for other dates can be calculated by addition or subtrac-tion, ii. 44. 1; *pl.* Rotes, ii. 44. 20.

Rowm, *adj.* roomy, large, wide, i. 2. 2. A.S. *rúm*, spacious.

Rytes, *sb. pl.* observances, ii. 4. 35.

Sadly, *adv.* carefully, steadily, with as little movement as possible, ii. 29. 12. W. *sad*, firm, steady.

Samples, *sb. pl.* examples, ii. 40. 4.

Sein, *gerund;* that is to sein, that is to say, *pr.* 25. *See* Seyen.

Semeth, *impers. vb.* it seems; me semeth, it seems to me, *pr.* 33.

Sen, *v.* to see, ii. 23. 27.

Senyth, *sb.* (1) the point of the horizon where a given azimuthal circle meets it; *hence,* the point of sunrise, ii. 31. 8; (2) the zenith, or

visible pole of the horizon, i. 18. 4. Arabic *al-samt*, a point of the hori-zon (shewing the *first* meaning to be the original one); whence Arabic *samt al-ras*, the zenith. *See* Azymuthz.

Septentrional, *adj.* northern, ii. 40. 30; *pl.* Septentrionalis, ii. 40. 28.

Sexe, six, ii. 42. 7.

Seyen, *gerund;* þat is to seyen, that is to say, i. 10. 2. *See* Sein.

Shaltow, *for* shalt thou, *pr.* 70; i. 7. 2.

Shewith, *pr. s.* appears (*used for the modern* is shewn), i. 7. 5; ii. 25. 4; 30. 6; 32. 3; Shewyth, ii. 26. 15.

Shipmen, *sb. pl.* sailors, ii. 31. 6.

Shrewe, *sb.* evil planet, planet of evil influence, ii. 4. 31.

Sin, *conj.* since, ii. 4. 3.

Sit, *pr. s.* (*for* Sitteth), is situate, ii. 7. 4; 37. 3. *See* Sitte.

Site, *sb.* position, situation, ii. 17. 24. Lat. *situs*.

Sithes, *sb. pl.* times, ii. 42. 6; Syþes, ii. 42. 7. A.S. *sið*, a path, a journey, a time.

Sitte, *pr. pl.* are placed, are set, i. 21. 6; *pres. part.* Sittinge, situate, i. 21. 8. *See* Sit.

Skale, *sb.* scale, or rather, double scale, for measuring both by *umbra recta* and *umbra versa*, i. 12. 2.

Slate, *sb.* a slate for writing upon, ii. 44. 3.

Slely, *adv.* slily, i. e. with great sleight or skill, skilfully, ii. 29. 13; Sleyly, ii. 29. 14.

Slen, *v.* to slay, *pr.* 44.

So þat, *conj.* provided that, ii. 29. 17.

Solsticioun, *sb.* the solstice, or point of the ecliptic most remote from the equator, i. 17. 5, 37. Lat. *solstitium*.

Sonne, *sb.* sun, *pr.* 55 ; used as a *feminine* noun, ii. 1. *rub.* A.S. *sunne*, G. *sonne*, Du. *zon*, Sw. *sol*, are all *feminine* nouns.

Sothly, *adv.* verily, soothly, *pr.* 15. A.S. *sóðlíce*, verily ; from *sóð*, sooth, truth.

Souereyn, *adj.* superior, ii. 28. 23. A technical term, applied to the western signs of the zodiac, as superior to the " obedient " eastern ones. *See* Obedient.

Sownyth, *pr. pl.* sounds as, i. e. means, i. 21. 37.

So3th, 1 *p. s. pt.* sought, ii. 45. 10.

Sper, *sb.* sphere, globe, i. 17. 15 ; Spere, sphere, i. 17. 28.

Spryng, *sb.* first beginning, dawn, ii. 6. 4.

Squyres, *sb. pl.* measuring-rules, i. 12. 2. Shakespeare and Spenser have *squire ;* from O.F. *esquierre*, F. *équerre*, a square, measuring-rule, from Lat. *quadratus*.

Stande, *pr. s. subj.* may stand, happen to be, ii. 34. 2 ; *pr. s.* Stant, stands, is situate, ii. 29. 5.

Statutz, *sb. pl.* statutes, rules, *pr.* 68.

Stike, *imp. s.* stick, fasten by insertion, ii. 38. 5.

Stok, *sb.* a stump or block of wood, ii. 38. 4.

Stont, *pr. s.* stands, ii. 42*a*. 3. *See* Stant.

Strange, *adj.* not its own, ii. 19. 4. Every star has its *own* degrees in the equator and ecliptic, viz. the degrees in which a great circle passing through the star and through the N. and S. poles cuts these circles respectively.

Straunge, *adj.* not well-known, ii. 17. *rub.* A *strange* star is one that is not represented upon the Rete of the Astrolabe.

Streitnes, *sb.* narrowness, smallness, i. 21. 33.

Stremes, *sb. pl.* rays of the sun, i. 13. 3.

Streyneth, *pr. s.* holds together, compresses, i. 14. 4.

Strikes, *sb. pl.* strokes, lines, i. 19. 2 ; Strykes, long marks, i. 7. 6. G. *strich*.

Succedent, *sb.* a "succedent" house, ii. 4. 29. The *succedent* houses are the *second, fifth, eighth*, and *eleventh*, as these are *about to follow* the most important houses, which are the *first* (just ascending), the *fourth* (just coming to the nadir), the *seventh* (just descending), and the *tenth*, just coming to the meridian.

Succedith, *pr. s.* succeeds, follows, ii. 12. 27.

Suffisantly, *adv.* sufficiently, *pr.* 26.

Suffisaunt, *adj.* sufficient, sufficiently good, *pr.* 7.

Suffise, 3 *p. pl. imp.* (let them) suffice, *pr.* 20.

Superfice, *sb.* superficies, surface, i. 21. 25 ; in the superfice of, closely bordering upon, in the immediate neighbourhood of, i. 21. 19.

Superfluite, *sb.* superfluity, superabundance, *pr.* 30.

Swich, *adj.* such, *pr.* 32.

Table, *sb.* one of the thin plates on which almicanteras are engraved, ii. 21. 4; *pl.* Tables, plates, i. 14. 2; tablets, ii. 40. 18. "*Tables* [in the last sense] be made of leues of yuery, boxe, cyprus, & other stouffe, daubed with waxe to wrytte on ;" Hormani Vulgaria, leaf 81.

Tak, *imp. s.* know, accept as a result, ii. 25. 34.

Take, *pp.* taken, ii. 3. 43; Taken, ii. 3. 41.

Tarienge, *sb.* delay, ii. 25. 20.

Tau3th, 1 *p. s. pt.* taught, ii. 44. 24 ; Tha3the, ii. 44. 32.

Techyng, *pres. part.* teaching, shewing, pointing out, ii. 12. 14, A.S. *tǣcan*, to shew, Gk. δείκνυναι.

Thank, *sb. sing.* thanks, *pr.* 38. A.S. *þanc*, an acknowledgment of a favour.

Theorik, *sb.* theory, theoretical explanation, *pr.* 59.

Ther, *adv.* where, wherewith, ii. 29. 9.

Thikke-sterred, *adj.* thickly covered with stars, ii. 23. 1.

Thilke, *pron. pl.* those, i. 7. 5.

Thise, *pron. pl.* these, *pr.* 21.

Tho, *adv.* then, ii. 1. 8.

Tho, *pl.* those, i. 8. 4.

Thowmbe, *sb.* thumb, i. 1. 1.

Thridde, third, ii. 35. 3.

Tid, *sb.* time, hour, ii. 3. 10; Tyd, ii. 3. 12.

To, *adv.* too, ii. 25. 19; Too, besides, ii. 45. 14.

Too, *num.* two, ii. 42. 16.

Tornen, *v.* to turn, i. 21. 3.

Tortuos, *adj.* lit. tortuous, i. e. oblique, applied to the six signs of the zodiac (Capricorn to Gemini), which ascend most rapidly and obliquely, ii. 28. 19.

To-þridd, two-third; to-þridd parties, two-third parts, two thirds, ii. 41*b.* 7; Too-þridd, *ib.*

Towchieth, *pr. s.* touches, ii. 27. 3; Towchith, ii. 27. 6; Towcheth, ii. 28. 5.

Towre, *sb.* tower, ii. 41. 2; Tour, ii. 41. 3.

Tretis, *sb.* F. treatise, *pr.* 4; *pl.* Tretis, treatises, ii. 34. 9. A translation of Lat. *tractatus.*

Tropik, *sb.* the turning-point, a name for the solstitial points, i. 17. 8, 38.

Tropos, *sb.* a turning; but interpreted by Chaucer to mean "agayn-ward," i. e. backward, i. 17. 8. Gk. τρόπος, a turn.

Turet, *sb.* the eye in which the ring of the Astrolabe turned, i. 2. 1. Cotgrave has, "*Touret,* the little ring by which a Hawkes *lune* or *leash* is fastened unto the Jesses." See the note in Warton (Hist. E. P. ii. 315, ed. 1871), which seems to make the word equivalent to a *swivel.* Cotgrave gives "a drill" as another meaning, which clearly connects it with *tour,* a turn. It seems to mean both a ring which turns round, and an eye in which a ring can turn.

Twies, *adv.* twice, *pr.* 34; Twye, i. 16. 12.

Verray, *adj.* very, exact, true, *pr.* 61; Verre, exact, i. 12. 5.

Verreyli, *adv.* truly, exactly, ii. 3. 41.

Vmbra extensa, *or* recta, the lower part of the "skale;" Vmbra versa, the upper part of the same, or the part perpendicular to the "cross-line," i. 12. 6. See Fig. 1.

Vnknowe, *pp.* unknown, *pr.* 13.

Vnremevid, *pp.* unremoved, without (its) being moved, ii. 46. 21.

Vnstraunge, *adj.* wellknown, familiarly known, ii. 17. *rub.* The *unstrange* stars are those which are represented upon the Rete of the Astrolabe. *See* Determynat.

Vouche, *v.* to vouch; vouche sauf, to avouch as safe, to vouch-safe, grant, *pr.* 72. Cf. *William of Palerne,* ed. Skeat, l. 4152.

Vp, *prep.* upon, ii. 1. 2.

Vppere, *adv.* more upward, ii. 12. 13.

Vsurpe, 1 *p. s. pr.* usurp, claim, *pr.* 41.

Vs-ward; fro vs-ward, away from us, i. 17. 9; to vsward, towards vs, i. 17. 39.

Vulgar, *adj.* ii. 9. 3. The *day vulgar* is the length of the "artificial" day, with the durations of morning and evening twilight added to it.

Waite, *imp. s.* watch, look, observe, ii. 5. 11.

Waityng on, *pres. part.* observing, ii. 38. 11.

Webbe, a cobweb, i. 3. 4.

Wegge, *sb.* a wedge, i. 14. 3. A.S. *wecg.*

Wenest, 2 *p. s. pr.* expectest, ii. 3. 44.

Were, *pr. s. subj.* should be; *also*, would be, ii. 43. 7.

Weten, *v.* to know, ii. 44. 30.

Wex, *sb.* wax, ii. 40. 21.

Wexcde, 1 *p. s. pt.* waxed, coated with wax, ii. 40. 17. *See* Tables.

Wey, *sb.* (1) the sun's apparent way or path during a given day, ii. 30. 3 ; see l. 10; *also* (2) the sun's apparent path or annual course, i. 21. 30.

Weyere, *sb.* the "weigher," a translation of the Lat. *equator*, because it weighs equally the night and day, since the days and nights, at the equinoxes, are equal; i. 17. 16.

Whaite, *imp. s.* watch, observe, ii. 25. 21. *See* Waite.

What—þat, i. e. which, ii. 17. 14 ; 18. 2.

Whereas, *adv.* where that, where, ii. 31. 13.

Whir, *sb.* wire, thin metal rod, ii. 48. 5. A.S. *wír.* The word should be spelt *wir ;* the MS. spelling *whir* is faulty.

With-drawe, *imp. s.* subtract, ii. 44. 27; Wyth-drawe, ii. 45. 4; 1 *p. s. pt.* With-drowe, ii. 45. 7.

Wol, 1 *p. s. pr.* (I) will, *pr.* 28.

Wombe-side, *sb.* the front of the Astrolabe, i. 6. 6. See Fig. 2.

Wot, 1 *p. s. pr.* know, ii. 3. 50 ; *pr. s.* knows, *pr.* 25, 50.

Wreten, *pp.* written, ii. 44. 7.

Wrowhte, 1 *p. s. pt.* wrought, worked, ii. 3. 27; Wroȝth, was working, ii. 45. 13.

Wyte, *v.* to know, ii. 3. 26.

Ycleped, *pp.* called, ii. 39. 3. *See* Clepen.

Yif, *conj.* if, *pr.* 72.

Yit, *adv.* as yet, hitherto, *pr.* 20.

Ylike, *adv.* equally, ii. 26. 12.

Ynke, *sb.* ink, ii. 5. 12.

Zodia, *sb. pl.* beasts, i. 21. 36. Gk. pl. ζώδια, from ζώδιον, dimin. of ζῶον, a creature.

Zodiac, *sb.* zodiac, *pr.* 65. An imaginary belt in the heavens, of the breadth of 12°, along the middle of which runs the ecliptic. The Astrolabe only shewed the northern half of this belt; see note on p. 13. Named from the imaginary creatures formed by the constellations situate in it; from Gk. ζώδιον, dimin. of ζῶον, a living creature. See a drawing in the English Cyclopædia, Arts and Sciences, viii. 1054, which shews the figures of the animals in the zodiac as represented on a ceiling in the great temple of Denderah in Egypt, sculptured about 716 B.C. There is a beautiful copy of this sculpture, in white marble, in the Fitzwilliam Museum at Cambridge. The twelve "beasts" there shewn are *all identical with those which appear in a modern almanac.*

Ȝere, *sb.* year, ii. 44. 2 ; *pl.* Ȝeris, ii. 42. 6.

Ȝif, *conj.* if, *pr.* 35.

EARLY ENGLISH TEXT SOCIETY

THE Subscription to the Society, which constitutes full membership for private members and libraries, is £3. 3s. (U.S. and Canadian members $9.00) a year for the annual publications, due in advance on the 1st of JANUARY, and should be paid by Cheque, Postal Order, or Money Order made out to 'The Early English Text Society', to Mr. R. W. Burchfield, Hon. Secretary, Early English Text Society, 40 Walton Crescent, Oxford. Individual members of the Society are allowed, after consultation with the Secretary, to select other volumes of the Society's publications instead of those for the current year. The Society's Texts can also be purchased separately from the Publisher, Oxford University Press, through a bookseller, at the prices put after them in the List, or through the Secretary, by members only, for their own use, at a discount of 2d. in the shilling.

The Early English Text Society was founded in 1864 by Frederick James Furnivall, with the help of Richard Morris, Walter Skeat, and others, to bring the mass of unprinted Early English literature within the reach of students and provide sound texts from which the New English Dictionary could quote. In 1867 an Extra Series was started of texts already printed but not in satisfactory or readily obtainable editions.

In 1921 the Extra Series was discontinued and all the publications of 1921 and subsequent years have since been listed and numbered as part of the Original Series. Since 1921 nearly a hundred new volumes have been issued; and since 1957 alone more than a hundred volumes have been reprinted at a cost of £40,000.

In this prospectus the Original Series and Extra Series for the years 1867–1920 are amalgamated, so as to show all the publications of the Society in a single list. In 1963 the prices of all volumes down to O.S. 222, and still available, were increased by one half, and the prices of some texts after O.S. 222 were also increased, in order to obtain additional revenue for reprinting.

LIST OF PUBLICATIONS

Original Series, 1864–1968. Extra Series, 1867–1920

3

4

Other texts are in preparation including three further English versions of the Ancrene Riwle.

Supplementary Texts

The Society proposes to issue some Supplementary Texts from time to time as funds allow. These will be sent to members as part of the normal issue and will also be available to non-members at listed prices. The first of these, Supplementary Text 1, expected to appear early in 1970, will be *Non-Cycle Plays and Fragments*, ed. Norman Davis (about 42s.). This is a completely revised and re-set edition of the texts in Extra Series 104 with some additional pieces.

May 1968

Publisher: LONDON · THE OXFORD UNIVERSITY PRESS, ELY HOUSE, 37 DOVER ST., W. 1

Printed in Great Britain
by Amazon